Ryan Vachon

Science Videos

A User's Manual for Scientific Communication

 Springer

Ryan Vachon
Institute of Arctic and Alpine Research
University of Colorado, Boulder Earth Innitiatives
Boulder, CO, USA

ISBN 978-3-030-09892-6   ISBN 978-3-319-69512-9   (eBook)
https://doi.org/10.1007/978-3-319-69512-9

Printed on acid-free paper

This Springer imprint is published by the registered company Springer International Publishing AG part
of Springer Nature.
The registered company address is: Gewerbestrasse 11, 6330 Cham, Switzerland

Preface

If you have ever had a hunger to make your own movie, now is the time to chase that down. It is an undertaking that requires equipment, time, and intention; however, results can be very satisfying. Beyond personal gratification, films are powerful marketing tools, ways of communicating complex subjects, and means for unifying people around a cause. With its immense popularity, film is a no-brainer way for communicating and sharing technical content. Fortunately, modern computer applications and technologies have turned video making into easier and potentially inexpensive enterprises.

The keys to making good technical films begin with understanding what is possible under the limitations of budgets and equipment. From there, you can take informed action to purchase and use what is within your means. Your mind and creativity, coupled with these tools, carry great potential to produce very impactful films. This book is your instruction manual for accomplishing just that. Through the eyes of a filmmaking veteran, it delves into what is possible and for how much. Realistic expectations are set for what it takes to succeed while conserving time, effort, and resources. Each chapter lends thoughtful perspective and directions on the individual steps along the road from conception to film completion. Your curiosity into filmmaking may be a passing fancy or the seed for a life-long enterprise. Find out where it fits into your life and work scheme.

I would like to give special thanks to the helpful thoughts and efforts of Susan Owen and Daniel Zietlow.

Boulder, CO, USA   Ryan Vachon

Contents

Chapter 1
A Short Walk Through Film History

Modern filmmaking is the rapidly maturing partnership of time-tested methods and rapidly advancing technology. A craft that was purely film based has now shifted towards more deft and powerful digital platforms. Footage recorded onto memory cards can quickly be uploaded onto hard drives or the web, enabling filmmakers to work together from afar. These advances present opportunities for sharing information and telling stories in effective new ways, reaching diverse audiences, and marketing products with techniques that were unimaginable a couple of decades ago.

A Brief History of Film

Film is everywhere in our daily lives. No longer do we have to travel to the cinema to enjoy a movie. We now have the capability to stream hundreds, if not thousands, of movies, documentaries, or television shows right onto our computers and mobile devices. Furthermore, we aren't even limited to watching films produced by large production companies or the Hollywood elite anymore. Anyone with access to a camera and a social media account can share their video with the world. With how integrated film is nowadays, it's remarkable to think that less than 150 years ago, movies weren't a reality. Heck, modern photography was still relatively new, let alone the idea of stringing together a progression of similar, yet slightly different, still images to create the illusion of motion.

Who made the first ever film is debatable. Many suggest that an Englishman, Eadweard Muybridge, might have nabbed the prize. While today many may think of film as primarily a tool for entertainment, Muybridge was motivated not by entertainment but by science. He also may or may not have been motivated by a bet. You see in 1872, then California senator, tycoon, and racehorse owner, Leland Stanford approached Muybridge with a seemingly straightforward question: when a horse runs, does it ever have all four hooves off the ground? Since a horse's legs move

© Springer International Publishing AG, part of Springer Nature 2018
R. Vachon, *Science Videos*, https://doi.org/10.1007/978-3-319-69512-9_1

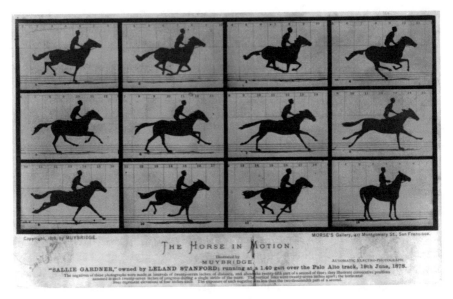

Image 1.1 Frames from the first moving picture, Eadweard Muybridge's *The Horse in Motion* (Reference: Wikipedia)

quite fast as they run, still photographs of the act were blurry. Stanford wanted Muybridge to develop technology to resolve the conundrum.

To tackle this question, Muybridge installed 12 cameras along Stanford's personal track. As a horse galloped, it would hit trip wires, thus triggering the cameras and producing 12 sequential photos of the horse moving down the track. Muybridge then spliced the photos together in the order that they were taken, such that when they were played in a manner called "stop action," the horse appeared to be running. One image settled Stanford's query: there is an incredibly short period of time when the horse was completely off the ground while running. With this question settled, Muybridge turned his efforts to producing several of these "stop action" films during the 1870s and 1880s. The first of these, *The Horse in Motion,* is often given the honor of the world's first film. While Leland Stanford went on to start Stanford University, Muybridge was at the cutting edge of an industry that has created something ubiquitous in households all over the planet: movies (Image 1.1).

By the 1890s, the film industry began to grow. The function of early films filled a very different niche from still photography. Film was an inexpensive way to give large numbers of people experiences that they could not otherwise afford – watching a live performance or traveling to an exotic land. Producers of films could bring actors, stages, and storylines to multiple cinemas around the planet simultaneously on any given day of the week! What's more, the functionality of film expanded to provide people with an exciting way to keep up to date on news and world events. Initially, these early films were short clips portraying one scene, in black and white, and without sound. This quickly expanded to encapsulate multiple scenes supported by orchestras, commentary by projectionists, and sounds to enhance the impact or authenticity of the film. People originally viewed films in primarily temporary spaces, but as this new medium of storytelling captivated larger and larger numbers of people, it ushered in the birth of cinemas. The Nickelodeon is heralded as the first theater dedicated to the projection of films, opening its doors in Pittsburgh in 1905.

During the early part of the twentieth century, the United States produced most of the films that hit global cinemas. Production companies in Hollywood became competitive, driving advances in filming and editing. Innovative lighting enhanced drama and humor. In 1927, soundtracks began to accompany moving visuals. Because of the competitive nature of production companies, it didn't take long for theaters all over the planet to start playing films with matching words, singing, sound effects, and background music, thus providing a consistently improved movie-watching experience at each screening. Audio was here to stay. The start of the Great Depression couldn't even put a damper on the rising film industry. Despite the widespread hardships, moviegoers kept visiting their local cinemas, arguably as a diversion from reality.

With the start of World War II, the function of movies in society expanded. In Britain, war dramas caught wing. The film industry also saw the rise of non-English language cinema during this time. Films from India, Japan, and a number of other countries proliferated. In turn, international producers started to steer popular movie themes and innovate filming techniques. Here in the United States, filmmaking continued to mature in its own right. Movies like *Hail the Conquering Hero* (1944) became a powerful tool for raising war time patriotism and propaganda. Films of jingoism spread their own problems. Following the end of World War II, many actors, writers, and directors were put under scrutiny in a large-scale Hollywood investigation for treason. In turn, this sparked an era of paranoia that influenced the themes of many movies, with aliens and foreign invasions driving many storylines. The loose portrayal of H. G. Wells, novel, *The War of the Worlds*, which came out in 1953, exemplifies this post war fervor (Image 1.2). In close pursuit, films began introducing questions about people's place in communities and whether society could be trusted. Thus, not only did we see film influence society in this period, but society and current events began to influence film.

In the 1950s, film went through a war of its own as cinemas became threatened by the arrival of another visual medium: television. Due to robust competition for viewer's attention, movie production studios and theaters started to close. To help retain viewership, filmmakers began to use increased screen sizes for more dramatic viewing experiences. Epic films, like *Spartacus* (1960), used the larger screen formats to increase their impact. 3-D films also emerged at this time, presumably in response to declining traffic to cinemas.

While the 1950s and 1960s represented a period of declining production in Hollywood, the 1970s emerged as a renaissance in film production. Coinciding with the Sexual Revolution, movie scenes and themes became more sexually and violently explicit, with pornography, as well as martial arts films (e.g., Bruce Lee's 1972 film *The Way of the Dragon*), catching greater wind. Before the 1970s, cinema often followed predictable storylines, but these new films preyed upon people's preset expectations of how a story should be told. Just look at productions like Stanley

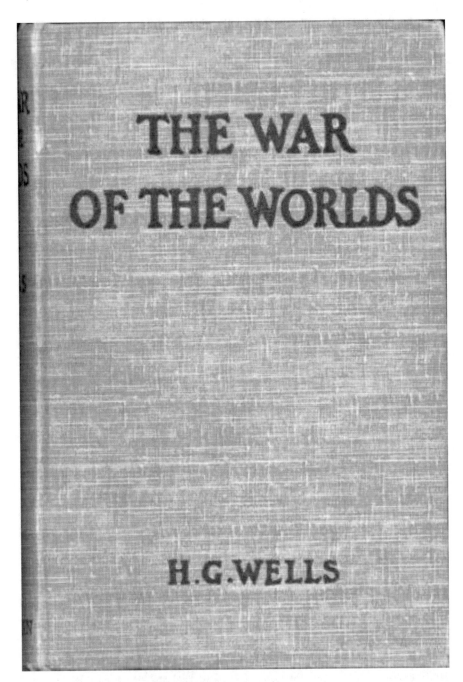

Image 1.2 The book *The War of the Worlds*

Image 1.3 A scene from *A Clockwork Orange*

Kubrick's *A Clockwork Orange* (1971), a film that shook people's movie conventions by telling a story full of strange events and an unconventional ending (Image 1.3).

In the 1980s, another cinematic war was brewing with the rise of movie rentals. The dawn of home videos was upon us. VCRs (and later DVDs and Blu-ray) meant that people could skip watching films in the cinema and just wait to watch it in their own homes. The advent of home viewing created ripples in traffic visiting cinemas. Advancements in technology weren't always a bad thing for the film industry, though, as the 1980s also saw developments in digital technology. Live-action scenes rapidly turned to visual effects for impact. The visual effects in *Terminator 2: Judgment Day* (1991) blew people's minds. In fact by 1995, the most successful feature-length story to be produced completely with computer animation, *Toy Story*, was completed.

Bubbling under the surface throughout the 1990s was another formidable revolution – streaming media, which became a mainstay for people wanting to watch movies by the start of the twenty-first century. The most obvious challenge to videos being broadcasted on the Internet was finding enough download bandwidth to play a video of acceptable resolution. These hurdles ignited an insurrection in video formatting. Companies have battled to develop faster data transmission over the Internet and more highly compressed films – all in the name of creating a more appealing market where people could enjoy watching film from their home theater or computer. Although the challenges are being tackled, streaming technology still struggles with meeting the viewing definition (quality) needs of demanding audiences and ever-increasing film resolution.

(Reference: Encyclopedia Britannica, www.filmsite.org, Wikipedia)

The Expansion of Independent Filmmakers: Including You

Technology has forever changed the way that professional production companies make movies. This high-end technology also streamlined the opportunities for most people, like you, to piece together compelling films. The combined impact of fast computer processors; inexpensive, yet capable video cameras; sophisticated editing programs; and Internet broadcasting (e.g., YouTube) has millions of people becoming independent video producers. Part and parcel with this conversation is a realization that millions of people are actually open to watching films at production qualities far lower than professional production companies. This is a sign of the times. While people's expectations for quality might be lower, the length of films that people are willing to watch has also dramatically decreased. The long tradition of canonical television shows lasting at least half an hour, or a movie lasting over 90 min has made room for films lasting no longer than 3–5 min.

The number of capable producers, paired with rapidly advancing technology, has positioned Internet-based films as the backbone for many businesses. Without big, fancy production companies, you can market your own product. Describe steps for assembly. Share a vacation. Tell the world why your science is revolutionary. Sell a house. Broadcast a series of video podcasts to share timely news. People are capitalizing on these uses every day because good video products can create huge followings.

For introductory filmmakers, this is the boon. Short films with decent production quality can have a profound impact on intended audiences. What's more, budgets for these films can remain quite reasonable. That said, the inundation of online films means that there is enormous competition for people's attention. How can we separate ourselves from other, possibly mediocre, filmmakers? Intention, knowledge, and skill can go a long way to elevate your film above the masses. Let's dive into creating impactful films.

Chapter 2
Know Your Audience

Responsibility lies on the shoulders of filmmakers to identify their intended audience, as well as understanding how this audience thinks. Knowing this, stories can be fashioned for optimal impact. Those who know what will elicit reactions from viewers are well-situated to craft a film that fulfills the proposed goals of the project. Gaining such understanding begins with researching the intended audience. The objective is to learn enough to empathize with their needs. These are the foundations of filmmaking. How footage and sound are combined into a final film is guided by these ideologies. Such sensitivities are best appreciated early in the moviemaking process, so time is not wasted and films are efficiently generated.

The Myth of Serendipity

Do you ever wish that everyone who watched your video saw its inherent value? Wouldn't it be nice if, after discovering your video's significance, these people then rewarded you with purchases, grants, or viewership? Indeed some work is very relevant or charismatic, meaning that large numbers of people are naturally drawn to investing in or learning about are market or field of research. Experimentations with time travel is an example that comes to mind. Proven moneymaking ventures, another. The survival of baby seals wins most hearts. However for most, we have to work hard to share our knowledge with the people outside our inner circle. This means that if we have goals to reach new audiences, we have to contemplate how we reach out. We have to know the needs of our intended audience.

Are you one who wants to make a film, post it onto the Internet, and hope that it will gain a massive following? Perhaps you visit *YouTube*, where the front page is graced by poorly composed films captured by unsteady hands. You may also note that some of these videos have tens of millions of views! These observations can give you the sense that a lot of people watch mediocre films. The truth is, yes, this does happen but these outcomes are often serendipitous.

© Springer International Publishing AG, part of Springer Nature 2018
R. Vachon, *Science Videos*, https://doi.org/10.1007/978-3-319-69512-9_2

Addressing serendipity: I suggest that no one holds back from submitting short films to places like *YouTube*, *Vimeo*, or *ScienceStage*; however, I think that anticipating great outcomes (i.e., lots of viewers) from this approach is a slippery slope. Let's put this in perspective. In 2014, over 300 h of video content was uploaded onto *YouTube* every minute (*Reference: tubularinsights.com*)! Three hundred hours of content every minute means 432,000 h of content every day. If an average video is 4 min, 20 s (0.07 h) long, this is like six million videos being uploaded every day. What you see on the front page of *YouTube* is a tiny fraction of all that is out there. These large numbers reduce the chances of a windfall in video viewership to those of holding winning lottery tickets. For most, leaning on such numbers is not a good use of business energies.

Through another lens, web platforms such as *YouTube*, *Vimeo*, or *ScienceStage* carry enormous volumes of videos and reflect the huge number of people engaging in this medium planet-wide. It highlights the public's openness for turning to video for entertainment and education through these means. Additionally, it unlocks a clear trend in people's decision-making when choosing which films to watch – potential viewers recognize films that have been watched numerous times, are intrigued by what drew others in, and thus give these videos a watch. The number of views skyrockets. Imagine if you could tap into this volume of traffic to meet your film-communication needs. If you are tempted to explore these dissemination networks, improving the quality of your videos will increase impact and the chances of gaining many views.

Outreach: Action to Inform

Outreach is a broad term for activities that infuse other's lives with knowledge, perspectives, and stimulus for action – usually around a dedicated theme. Outreach is how we market devices, show how to use medication, or communicate instructions.

Outreach also includes the labor that goes into increasing the literacy of a certain audience. Literacy is much more than a snapshot of an individual's level of knowledge. It encapsulates a person's disposition towards information and how they speak about the content. It is guided by their experiences and learning (to date) that brought them to a belief that a field of work or piece of information is either important or tangential. It defines whether they are open or resistant to learning more.

With these thoughts in mind, our transmission of information will be filtered through our target audience's curiosities, biases, interests, and decision-making. If we realize this, we can create positive experiences for these people and thus initiate a self-sustained landslide of reflections. Because most business activities want action or interaction with others, the hope is that the reflections result in voting, investment, partnerships, or additional research.

A Communication Exercise: Establishing Rapport

The methods for how we reach general audiences are very different from those that we use to reach fellow technical specialists. Experts in a given field frequently want interactions to cut to critical information, like the details of results. Conversely, novices or newbies in the same field may more details in the buildup to critical information. In many situations, the latter are the cohorts of people that we want to reach. They represent the greatest potential for attracting new minds, investors, or action.

Perhaps, deep in your heart you get that carefully crafted dialog, and a well-paced flow of media are effective for impacting viewers. Knowing is one thing, yet manifesting takes conscious effort and a rich understanding of what works to these ends.

A good entry into storytelling begins with building a rapport with your audience. This example might warm you up. You are at a dinner where the people nearest to you are strangers. A person across the table from you inquires about what you do. Being put on the spot demands a reaction. One response may be to clam up and hope that the person leaves you alone. More often than not, they will leave you alone, with the entire corner of the table falling into awkward silence.

Oppositely, in this situation many of us veer towards answering this question with some level of positive engagement. Deep in the back of our brains, perhaps even unconsciously, we have a couple of goals that steer our approach. For me, they might be being friendly and facilitating easy conversation for a good fraction of the dinner. Perhaps my description of what I do will prompt the individual to ask questions that lead discussion to reveal their interests in science or related knowledge, or they might share interesting information or experiences about their life. Wouldn't it be nice for the night to carry on and both of you come away happy that your minds have been expanded?

How do you execute this conversation? What do you take into account as you devise what is most useful to say?

A research scientist, graduate student, lab manager, or product developer shouldn't likely jump headlong into the complicated thoughts that guide their daily activities or large-scale projects. For some, it has taken them decades to foster methods, knowledge, and, most importantly, enthusiasm for connecting the dots between very complex or science-rich concepts. To serve our goals at the dinner table, we formulate a dialog much like a highlight reel (a condensed video highlighting the tone, relevance, and themes behind a larger story). It is an account of what you want the person to understand in a short amount time, what is a gripping way to portray the concepts, and how much information you think that they can juggle while still enjoying the conversation. As you progress, you watch them for cues indicating that you rushed through information or skipped a critical link between one theme and another. You are keyed in on the truth that the conversation will crumble if you don't keep them on the hook – leaving your end of the table cloaked in silence.

The skill of devising conversation that balances your goals with an audience's needs is essential for all good storytelling.

At the foundation of the dinner party example is a mentality that can help us build stories for audiences of all backgrounds and knowledge levels. Engagement hinges on understanding that people's knowledge, interests, experiences, and mood are individual. In order for us to affect people deeply, we are best served when we appeal to these elements of people's lives (Image 2.1).

This model of thinking expands to more significant occasions, like during business-related pitches. In more professional instances, I have taken a moment to remind myself that dinner party approaches still apply here. Just because my mission is to share science or technical content with very structured goals in mind (rather than fill in space at a dinner party) doesn't mean that thoughtful engagement is thrown out the window. The same tenet stands – if I share too much information or make conceptual leaps beyond my audience's capacity, I lose connectivity. Over an informal

Personal Anecdote

My earliest memories of learning about physics did not blossom from formal schooling, but rather from late nights sitting down with my dad near the wood-burning stove, scribbling notes on crinkled paper napkins, discussing how to design motorized surfboards. Or we would flesh out the reflective qualities of white materials over shiny. I was 12 or 13, but my father would give me the smallest of glasses of wine to sip and ponder. I felt adult and warmly invited into my father's world of mechanical engineering.

To this day I think of him as the go-to guy for using the equations of Newtonian physics to describe how our mechanistic and natural worlds function. Napkin after napkin, we would make rough calculations and design cool devices. I was entranced by the idea that our minds could solve practical problems. He would then show how numbers could lead to saving money and time, or improve performance. A little thinking could pay off, literally. I am guessing that his primary mission was for me to have healthy, "controlled" first drinking experiences. However, he leveraged his command of engineering principles to get me to think quantitatively. He later told me that he would put himself into his own shoes as a kid, getting drawn into engineering through ham radios, and projected them onto me. The upshot? Long before I knew what the word "physics" meant, I learned to use force diagrams to add immense depth to my everyday experiences.

Indeed, I was a kid at the time that I was sipping wine and drawing on napkins. While it might not be appropriate to use this teaching style for all kids, it was very effective for me. It is a broader lesson in building great learning experiences. When done correctly, people can be magnetized to notions that are outside the norm. The foundations of teaching and engagement begin with putting ourselves into the emotional and intellectual shoes of our audience (Image 2.2).

INTEREST IN CONTENT

Knowldege

Experience So Far → Traction

Mood of The Day

Image 2.1 Numerous personal factors play into how people engage issues and concepts. The more aspects that you can appeal to, the better chance your work will gain traction

Image 2.2 As a kid I learned physics and how to drink wine like an adult, at the same time

Box Models

Physics

Reflectivity

dinner, the value of this effort might not feel perilous, but when it comes down to pitching our work to a purposeful crowd, impact is everything.

Communication Theory: The Art of Empathy

Sharing information is largely about creating an experience with which the audience can comfortably connect. If you lose your audience for a second, they might not make key associations for which you had hoped. As a storyteller, your most powerful tool is to understand the needs of your audience so that you can infuse their lives with your vision. In academic circles, taking account of a user's needs is integral to *communication theory*.

Communication theory was born out of math. The initial question was, how could a user reproduce numbers in a place or application different from its original use? More specifically, *communication theory* delves into the communicator's ability to read and regurgitate symbols and meaning from the origin and embed them elsewhere. With numbers, the reproducibility of information is measurable; however when we are dealing with the transfer of information between people, figuring out how much information is retained and lost is quite challenging. Why? Information has elements of subjectivity. The missions of some acts of communication affect or transmit wisdom, perspective, emotions, and bodies of thought. Similar to talking at a dinner party, psychology and sociology are critical components of *communication theory.*

Blossoming from *communication theory* are advanced methods and theories for streamlining information and concept-transfer between people or groups. People devote entire professions to studying and developing *communication theory,* and identify its prized value in building impactful relationships, pitches, and learning devices. For the purposes of this book, we will highlight a few key notions at the roots of this field. Indeed, these begin with maximizing the potential for your transmissions to be received.

The process of impactful communication begins with listening to the people that you wish to reach. What are their values? How much time do they devote to information versus how a situation makes them feel? What are their emotional needs? Empathy is the name of the game. Empathetic communication often opens individuals to more fruitful engagement with media, particularly regarding topics around which strong sentiments revolve. It is a fast way to diffuse feelings of weakness or shots to the ego when people are asked to learn or change their opinions. An otherwise threatening experience can be repurposed into one that is sharing or nurturing.

Empathy might not mean that you share the exact same experiences, so much as you can identify situations in your life that brought you to feel similarly. Ponder how a scenario, person, or product paved a pathway to a personal realization. This is an application of looking back and owning where you once were at the time that you felt similarly. More often than not, this gives you a hook to begin engaging other people with the information you wish to present. We want to examine the hooks that make learning irresistible, an inspirational experience, or a trend so vile (or amazing) that you needed to act. This is first-hand confirmation of successful methods that you can draw upon to create the impact that you desire. Once you have recognized that you *get* your audience, you can use yourself as a resource to guide construction of your video and filter pacing, tone, and level of information that will help to change your audience's level of literacy.

At first, it is helpful to formalize the self-inquiry process. I would suggest that after you identify your intended audience, set aside dedicated time to think about these questions. Write your answers down. Let's go through an example. As part of this section, I will share my own answers to how I was able to begin investing my life savings into the stock market. The outcome of my example was not the viewing of a film; however the same principles stand for practically all situations where new audience engagement is the goal.

- What was the experience that left you feeling like your intended audience? (*I was intimidated by the uncertainty of the stock market. How could I make choices when I was busy and didn't have time to make educated investments?*)
- How resistant were you to learning the content and what do you perceive created this resistance? (*I was paralyzed to think that I could be the cause of losing my life savings.*)
- What did you need in that situation to grasp your full attention? This is a little different than answering "what drew you in." Again, this is an exercise designed to put you into people's shoes, including your own. (*I didn't have to know everything to start investing in the stock market. I learned that I could start small and work with people who understand the system and can show you how to make safer investments.*)
- What were the hurdles that the makers of your experience needed to overcome to access your mind? And how did they accomplish the task? (*My gridlock was freed up by getting me to try a small investment under the guidance of an expert. They asked me to be given a chance to help me with a small sum of money.*)
- What were a few things that the facilitators did to change your emotional disposition towards engagement? (*He empathized with my fear that I could lose lots of money. What's more, they appealed to my need to start my learning process in small doses.*)

Second, figure out how you can create a similar experience for your audience. It is preferable to answer these questions immediately after you answer the first set.

- What do you think your audience will be feeling before they engage your ideas? (*Gripped with fear*)
- If you were in their shoes, what experience do you think that they would need to connect with the content in the desired way? Understanding? Information? (*Patience, gradual exposure to success, and understanding that I was intimidated*)
- What are three to five mechanisms that you can use to create the transformation for which you are hoping? A mechanism could be giving a nod to people's frustration with the current state of knowledge or showing, in vivid imagery and testimonials, the outcomes of certain steps. (*My friend talked about how he had to make mistakes with small sums of money before he learned that a diverse portfolio protected him against market fluctuations.*)
- What would be too much and not enough information? (*The fundaments were all that I could handle at first, and then checking with me on whether I was ready to graduate to the next level of investing.*)
- What are a few questions that you can ask of others (who might have insight) to help guide you to increased connectivity with your audience? (*What methods have you used to break paralyzing fears?*)

This practice simplifies and sharpens your message. You will find that certain information rises to the surface as your stories are guided by empathy, while other information, that you once thought was important, stands a good chance of becoming distractions and overwhelming. A necessary step is owning such reductions or disruptions, and culling these elements from your design.

As part of my own exercises in empathy, I like to approximate people's dispositions regarding certain factors. By putting a number value to certain questions, say on a scale of 1 to 10, I can guide how much effort I ought to put into accomplishing different goals of my films. For example, I often ask, "What are the knowledge, interest level, and learning potential of the audience?" Let's say that the goal is to educate second graders on new science ideas, such as water being used in the manufacturing of cars. They might know very little (so, a 2 on the knowledge scale), have great interest (an 8 on the interest scale), but have very few tools to understand complexity (a 3 on the learning potential scale). For this, I have to throttle back my learning expectations, yet realize that simplifications of complex lessons could shape a rich appreciation for some of the primary principals.

Without understanding the needs and inclinations of your audience, elements of your stories become fragmented, and it is more difficult for audiences to come away with a single lasting impression or lesson learned. Understanding the needs of your audience means that you can concoct the glue to keep all of the moving parts of your films together. The stronger the glue, the more easily a viewer is transported from one topic to the next and the easier it is for you to achieve the goals of your film.

Chapter 3
Storytelling

Next, it is up to you to build your own story. This step benefits from a rich understanding of storytelling methodology. What's more, there are well-known tools, such as writing a *storyboard*, that help formalize the steps that you will take to film and edit your movie. Early planning efforts and organization of your media will pay dividends in the long run.

> **Personal Anecdote**
> One morning I was working at my computer in a local coffee shop when I ran into a former student of mine. I am friends with this young woman and am somewhat versed in her research. However, I am not naturally drawn to her specific field of work. Our conversation turned to her recent submission of her honors thesis. Overjoyed, I congratulated her, but in the background I was picking up on her wanting me to read her thesis. Little red lights went off in my head. I immediately identified an imbalance between the effort that it would take to read 58 pages of mostly technical text and my interest in her work. I firmly believe that her work is important, but in a world overflowing with information and distractions from my personal goals, I want to be careful about what commitments I engage.
>
> The student used a great sales pitch! She then told me that she thought of me while she was writing her thesis. The mapping tools that she applied in her research produced some interesting animated graphics showing how specific data, say human population, relate to geography. She then told me that I could read her thesis online with examples of the animated figures, if I like. Her tactic of turning her thesis into a possible tool for my own applications turned my head, and I was drawn towards and, in fact, did read the document.
>
> Just like this woman's thesis, filmmakers need to create simple hooks for captivating viewers into greater personal engagement. As we pitch our work to other individuals, the more intriguing, compelling, and personally relevant we can make our concepts or outcomes, the more heads we can turn.

R. Vachon, *Science Videos*, https://doi.org/10.1007/978-3-319-69512-9_3

What Works for One May Not Work for Others

No matter how perfect, novel, and tractive the concept of your film may seem to you, it will not resonate with everyone. Many new filmmakers operate under the misunderstanding that their passions, interests, and deep insights will make for a universally impactful film. Some of the films of which I am most proud, and that receive great reviews among some audiences, disappoint or do not impress others. As filmmakers, we have to be okay with this: not pleasing 100% of the people 100% of the time. This reality highlights the importance of understanding the needs of your audience, as we discussed in the previous chapter. In turn, this can help us set realistic expectations for the impact of our work.

I have suffered from profound disappointment when I think that I have identified a silver bullet for conveying a story or body of knowledge. My first inclination is to direct this enormous head of steam into filming and editing my vision. Time and time again, I have learned that a wee bit of restraint leads to a better film. Sure, a great idea could lead to a spectacular connection between the film's content and my audience; however such enthusiasm runs the risk of breaking the rules of empathy. I like to ask myself if my personal associations to the information will lead to discontinuities with my audience. Are my individual interests leading me to unrealistically believe that others will find them similarly fascinating? Perhaps I can identify this happening on my own, but I frequently go to friends with my thoughts. Half an hour of well-directed questions has saved me days of effort.

Now that we have given a nod to people's individuality and reemphasized the importance of knowing our audience, let's go about figuring out how to tell your story. Since the vision is yours, the best place to start is with an exercise to formalize what that concept looks like. This is a great staging place for strategizing the tightest story that will accomplish your goals with the resources that you have.

Jedi Mind Tricks: How to Set Mental Groundwork for a Successful Film

One of the greatest skills for making films is training your brain to work in a mechanistic manner. Structure reduces the chaos that can ensue when undertaking large, multifaceted projects, and serves to guide me through the numerous steps that more reliably produce a great film. In the end, this saves time and resources. A first step in adhering to these mechanistic methods is to ask myself a series of questions (discussed below) at the start of every film project. These guide me towards answering *what is the goal of my film?*

What Information Will Your Video Convey?

In all likelihood, you have a certain amount of expertise in the subject matter of your film. This knowledge informs what you want your viewers to learn or experience. Because of this expertise, I find it very easy to slip into a mode of wanting to share as much knowledge with the audience as possible. If audiences are exposed to too much new information too quickly, they easily become disengaged from the video. To counteract this, it is helpful to take a step back and create a simple, one sentence mission statement for the video. "I want my audiences to understand why DNA is a double helix." "I want my audience to appreciate the critical role that drought plays in social unrest around the planet." The story that you build should revolve around giving this singular goal traction. I suggest writing your mission down on a sticky note and putting it in a place where you will often see it. That note is mantra! It will keep you directed and focused as you delve into different activities and minutia of building your story.

Who Do You Want to Reach?

As we already discussed, identifying your intended audience is of great importance. Are you making this for peers, policy makers, a film festival, your local AARP, or 7th graders? Knowing this will help you determine the level of detail, tone, and pacing that will transport your audiences to enlightenment. It is helpful to determine the audience you want to reach (possibly seeking advice from those who have more experience in this terrain) early in the process of planning your film production. It would be terrible to build a fantastic animation and then learn that it isn't the best approach or tool for reaching your greater goal. As with all efforts, enthusiasm rapidly dwindles when we catch ourselves spinning our wheels or loosing time with impractical efforts.

What Are the Impact and Tone for Which You Are Hoping?

Impact means that you want change to affect your audience. What change are you hoping for? What action do you want the audience to partake in after viewing the video? Sometimes identifying the desired impact helps to identify the filmmaking techniques that will lead you to your end goal. Later in the book, you will learn that certain filming methods and resources are more often used when building, say, a documentary versus a cartoon. Knowing what you want for your audience is crucial for identifying the style of your film.

Tone goes hand in hand with knowing the desired impact. Tone describes the feel of your video, and plays a large role in dictating how audience members react to your video. Maybe these words will help you identify what tone you want: *accomplishment*, *mystery*, *curiosity*, *wonder*, *inspiration*, *comedy*, *cutting*, or *instructional*. *Wonder* might work when sharing a philosophy, *inspiration* seeds sports action, and *instructional* suites teaching methods. I like to remind myself that a person watching any video will identify a tone. The question is can I build the tone that I want for them and thus transport them one step closer to my goals? You will later learn that filmmakers evoke certain tones through music, color characteristics, narration, types of scene transitions, and the speed that clips change. These are invaluable tools for producing a specific feel that you are seeking.

Cool Trick

Tone can change throughout a film. Think about how films present a problem and then how the filmmaker brings resolution to that problem. Both components of the story can be strengthened by differing tenor. Feature-length movies leverage this to transition despair to hope. Confusion to realization. Be aware that switching tone in the middle of a film is very powerful in productive and destructive ways. Misplaced tweaks in tone can leave a viewer disconnected. I have been challenged again and again with these issues when communicating the climate sciences. Since climate is surrounded by fragile feelings in some of the public's eyes, a small slip from an informative to didactic tone loses viewers in a heartbeat.

Model Success: Don't Reinvent the Wheel

As you set the mental groundwork for your film and finalize the information you wish to convey, the target audience and intended impact, you will soon have to identify the media, such as cartoons or expert interviews, to carry your story. This is intimidating yet is a large step in turning your vision into a reality. A great place to start is to watch other's work. There are great opportunities and logic behind using tried and tested methods in storytelling and filmmaking. Progress is more efficient when we travel a path that has, at least in part, been previously established.

Filmmaking is rarely about breaking every rule for the most unique viewing experience possible for a viewer. Using what works is not a commentary on your lack of vision. It is practical. One way to discover what works is to use the work of other people as a model for how you might reach success. See if you can command the tools that the person, used. It is like you are working from a predefined master template, yet your ingenuity and unique information will create a distinctive piece. This is not referring to plagiarism of other's work, but instead appreciating reproducible methods.

When I am lost for how I should approach producing a film, I go on a hunt, seeking other films that impact me in the way that I wish to impact others. For short films, I begin by looking at videos online. For longer films, I scour Netflix and iTunes. I talk to other people about my concept and inquire about whether they know of other films that might be similar.

From these searches, hopefully one or two films stand head and shoulders above the rest. Learn from these films! Dissect them! Remember, in this process you are looking for the elements of an equation that made the film resonate with you. Figuring out this equation is not always easy, the process benefits from some structured inquiry.

I usually undertake a formal exercise (discussed below) immediately after viewing a possible "template film." Even 5 min of distracted thought can skew my perception of why and how a film impacted me. I write my answers down, as research shows that putting thoughts into written words make abstract thoughts more tangible. What's more, it provides notes that I can look upon later on in the filmmaking process. Here are some questions that might guide your efforts.

What Is the Style of the Film?

What style of film do the filmmakers use to engage the audience with their mission? Identifying the film type opens our eyes to the ways that we can build our own film. When most people think about the types of film, they jump to their own experiences with Hollywood films: sci-fi, action, epic, western, horror, fantasy, love story, film noir, and comedy come to mind. This is a good start in identifying how films can be categorized; however these are all film genres, not film types. Film type is a discussion about the methods and media used to construct a film. Are you going to produce a cartoon or documentary? Once you recognize these qualities, you can start to identify the resources and methods that will likely facilitate the production of your film.

Film Types

- *Feature length films* are films with running times over 40 min.
- *Short format films* are often described as anything that is shorter than a feature-length film. While there is no consensus on the longest length of these short format films, people generally consider them less than 40 min.
- *Narrative films* are a type of fictional film (movies that have imaginary events or characters). The mission of a narrative film is to use events and characters to portray the story in ways that could be construed as real.
- *Documentaries* are non fictional films that describe one or several aspects of a historical event or ongoing situation. They are meant to depict a story based on real events in educational, yet often times, inspirational ways.
- *Music videos* are often short format films, because their length is limited by the duration of the song that drives the visuals. Typically, these films promote a song, however may also market elements shown or introduced during the song.

- *Photo or video collages* are driven by matching music to visuals. They are different from *music videos* because they are not limited to one song.
- *Animation* is an illusion of movement created by a sequence of non-photographic images with slight differences in position and geometry of scene elements. (positions and geometries of scene elements), When the images are projected rapidly in succession, elements within the images appear to move. Applications include cartoons, computer-generated imagery (CGI), and infographics.

The terminology used in describing film types is useful when sharing your filmmaking efforts with other people. Such language quickly aligns their understanding of your film with what you are hoping to accomplish. Plus these terms can dial in your search for more examples and learning resources.

Does the Style Suit Your Needs?

Review films that can most help you towards your goals. Sure, a certain film might make you stop and remark, but does its style align with your filmmaking options. What if you were drawn to one of the Harry Potter films, because it nurtured the development of characters over 90 min or a series of movies. Such monumental filming efforts might not mesh with your time, resources, and financial budgets, thus rendering the film style harder to emulate or learn from.

What Medium Was Used to Present the Core Concepts or Plot Progression?

Behind film types is the amalgamation of media. Interviews? Graphs? Photographs? Shots from fieldwork? A blend of all? For example, modern space exploration movies often lean heavily on cartoon or high-end computer-generated animations. Once you identify the medium, you can begin to wrap your brain arounding how you can procure the shots that will lead to your story to unfolding compellingly and logically. As part of the brainstorming process, consider listing the means at your fingertips for producing or harvesting this medium.

How Does the Film Draw You In to Keep Watching?

What tantalized you about the film? What was the entry point into your heart and mind? Take note on what tools they used to reach those ends. Was it very clear explanations or exquisite visuals? Perhaps the topic is new to you and the film used funny anecdotes as the hook. Some films ask open-ended questions of the viewer. Some act like a teacher and explain the importance of the subject matter. Many longer films present the storyline with a cliffhanger where the ending is not clear until you watch it to completion. Often times, there is more than one method that makes the film stand out. What were the devices that gave it traction? Music? Transition types? Color? The tone of narration? Take note about how subtleties and qualities "make or break" those segments of the film. Paying attention to all these questions help to identify methods to draw in your audience.

Often, you may find that it's a combination of elements that produces a high impact in a film. It is empowering to take a moment to think about how the entangling of different factors produced a great end result. The factors that produced the greatest outcomes may not be what you originally supposed. For example, perhaps you thought it was the black-and-white color scheme that sent shivers of fear down your spine, when instead it was the faraway shot of the shadowy figure. With realizations such as this, you are topping off your toolkit with wisdom.

What if it was the entire storyline that held you spellbound, rather than interesting isolated acts or segments? In these cases, take note of how the filmmaker built their story arc (or progression of events within a story). When did they present information, tension, and resolution? How rapidly did they move through each phase? One may be surprised about how rapidly the concluding point of a film unfolds when compared to all of the plot or background that built you up for the finale. One can further break down the investigation of a story arc into the stages of the film that you liked (or disliked) most. As you note when the film accomplished certain elements, like finding resolution to a great challenge, you can observe how certain scenes were constructed for the intended impact. Again, note what tricks or tools were used to accomplish this. Figure out what it was about these mechanisms that made them effective. Then do it yourself!

How Were You Made Aware of This Video?

As important as the making of the film, your film has to be injected into the lives of your intended audience. What was the pathway that landed the film that you like on your front doorstep? Streaming? A film festival? Be a critic. Was that an effective way to get it done? If so, can you consider doing the same thing? If not, how could you do it better?

The Art of Storytelling

A good video draws you into a world that the filmmaker designed for you. Some of the most compelling videos tell stories effortlessly, and you come away from a screening (or viewing) invigorated, interested in the content, and/or charged with new awareness. Just like a finely fashioned article or book, the ease by which a film engages viewers is not a measure of the time that it took for you to make the film. In fact, I argue that it is often the opposite.

Simplicity is an art form. I might understand the background to a field of science exceptionally well and have intense interest in its outcomes, yet developing a story that conveys that knowledge, fascination, and insight takes immense effort. Much like the values of early and concise organizational efforts to define your goals, carefully develop your storyline. This will amount to a better product and earlier completion. I firmly believe that without upfront efforts, you are doing a disservice to the desired outcomes of your intended film. Early efforts may seem overly controlling and perhaps the romance in filmmaking, as a creative outlet, may be lost. However, as skill with storytelling becomes second nature, concerns will likely fade.

Where to Start with Storytelling

Our lives have been imbued with storytelling from early on. Perhaps we learned that we could fib our way through a mistake. For our fiction to work, we had to fabricate a convincing reality. We've also laughed for an entire coffee break because the intern had a great tale to tell.

Both through words that we weave or listening in on how other people relate stories, we know that some stories work for us, yet certain blunders will immediately disengage an audience. A lot of us can list what people should NOT do when telling a story. What are some that come to mind? Never tell the ending prematurely. This rule stands for all friends watching a movie, and also applies to filmmakers who may have a perfectly formulated finale, yet diffuse it by leaking or dragging out the ending.

How about a couple other examples? Dull personalities are rarely good storytellers. Or, don't skimp on expression or the gory details.

Identifying what works for telling stories is equally as important, however more challenging. Several key variables that appear to be integral to successful stories have emerged from reading, attentive listening to professional storytellers, and my own thoughts on the process.

Conflict

You need conflict to drive a tractive narrative. Many Hollywood films portray characters that are confronted with a problem that they must overcome. Or, the advancement of a field has been beleaguered by difficulty, yet the innovation of a team breaks the gridlock. The hope is that a viewer can connect with the problem and realize that one product is here to help them. In this last case, storytellers are capitalizing on a pre-existing conflict in the lives of their audience.

In order for changing conditions to ignite interest, a storyteller must create conflict potential. Perhaps think about it like energy – in order to unleash kinetic (action) energy, there has to be potential energy. The rock must be on top of a hill (potential energy) in order for it to start rolling (kinetic energy). Without tension, there would not be a need for resolution. In this mind-set, easy-to-resolve conflicts present small potential for an impactful story. Good storytellers are masters of portraying conflict in the most vivid, personal, and compelling ways, paving way for the most exciting and transformative plot possible. If you want to get technical, the progression of events to resolution is the *setup*, *buildup*, and *payoff*. Collectively, this is called a *story arc*.

Background

Every story must be built upon a foundation of information or events. This is often referred to as the *backstory* or the *background*. Build a description to ground the audience to relatable experiences. Even if the story is fiction, the audience will create their own connections to far-fetched ideas through personally relevant analogies or their imagination. Backstory establishes the stage for a tension-filled plot.

Characters

Developing characters is fundamental to good storytelling. In the simplest interpretation of storytelling, characters are the main elements of the film with which the audience relates. Indeed this suggests that characters don't have to be people. Heck, they can be talking animated question marks. Characters, like the background, benefit from careful and vibrant portrayal. Indeed, character traits can foreshadow how the plot will unfold and even help build tension. Characters are typically relatable, yet unique and quirky. This makes them memorable and authentic. On a less literal note, the viewer can fill themselves in as a main character. Let's say that your story transitions through a series of beautiful landscape shots. The viewer's feelings and thoughts about the scenery will help to build your desired conflict and transformation.

Focused Content

Overly complex storylines can take on a life outside of your core mission. Thus, filmmakers must focus their attention on using content that is productive, not distracting. Too much information will leave the primary storyline diffuse, unexciting, or overwhelming. Not enough information leaves viewers disconnected from the continuity of events or logic behind innovation. Pay close attention to the purpose of all story elements that are incorporated into your film. For example, in order to establish a healthy background to America's spectator sport, baseball, you may wish to portray the development of spectator sports in general. Sure, battles within the coliseum were part of the history of spectator sports; however perhaps keep the details about how chariots were constructed to a minimum, when moving on to the British game, *rounders*, would be more productive.

Pacing

The pacing and tone of films are wonderful ways to evoke feelings in an audience. With time filming and editing, you will find that you can partially control an audience's perception of a situation through how long clips last, framing of shots, clip transition types (such as fading to black), music, and chromatographic edits. As such, potent storytelling through film also means that you can accurately weave these elements to steer your viewer's screening experience.

Note Before you decide that technical films might not need such variables, think again. Even instructional videos tell a story and benefit from attention to how the story is woven together.

Navigating a Science-Rich Story Like a Publication

One of the simplest ways to develop a storyline about technical content is to treat it like a science publication. As with many peer-reviewed papers or synthesis reports, films can be broken into an abstract, background, methods, results/findings, conclusions, and a summary. We will call each segment an "act". This structure is most useful for films designed for people who have interest in and knowledge of the content at the outset. In such cases, the intended viewers might be colleagues, scientists, or students.

Why would this structure of acts suit your needs? It is a simple template guiding a filmmaker to establish an audience's connection to results and findings in a manner that sheds valuable perspective on its contributions to a field. Executing the backstory and methods, and how they lead to the conclusions, is quite straightforward. What might you want to consider as you build the acts of such a film?

Title

It is always safe to begin with a title that embodies the notion or information that you hope to convey. It is common for new filmmakers to latch onto a title that is witty, profound, and flowery, but ultimately not all that relevant to the mission of the film. I don't suggest calling a film "The Curious Blue" if you are talking about how ocean buoys identify the migrating depth of the thermocline in the Central Pacific. Consider picking a title that is more literal. Even if you change the title later, a more direct title will serve as a continual reminder for what you are looking to accomplish with your film. I have caught myself cutting and pasting my mission statement into the title panel, knowing that during the filmmaking process I will come up with a more engaging title. If you are wed to the idea of a creative name, but fear that it might be over the top, consider putting the more literal name parenthetically following the more abstract one. Something like *The Curious Blue (How ocean buoys identify the migrating depth of the thermocline in the Central Pacific)*. Test both titles out on friends who are willing to review your work.

Abstract

The abstract is where you summarize the story that will unfold during film. In short words, share why your field of work is relevant (especially to your intended audience). Then go on to discuss why the information that define the background and the results of your work are relevant to this field. Share the take-home message or value of your product or work. Are you adding a new method, new results, or shaking the tree and adding skepticism to what is known? Give your audience the most condensed and compelling version of why they should keep watching and learning. You might think that this practice would go against the rule of storytelling (don't give away the story's ending). However, in this situation, it is usually not a surprising outcome that draws in viewers so much as the ways that you manifested fascinating determinations or results.

Background

Now that you have hooked the viewer into digging into the details, set the stage for where the information fits into a broader research, market, or field. What is the history that brings you to the question that you are answering? What is the chronology of technological advancements or theories that people must know in order to think critically about what you are introducing? Regardless of people's past awareness, sharing a healthy appreciation for where the work fits into the broader spectrum of a body of work is often evocative and lends credibility to research efforts.

Let's go a step further. Knowing your audience helps to put boundaries on how much information you can share. When my intended audience is poorly informed about the topic of a film that I am producing and I take them on a comprehensive tour of knowledge. I try to build an intuitive, simple, and progressive backgrounds to my film. If the audience misses one link in the chain of understanding, I stand a chance of not accomplishing my primary goal.

For very informed audiences, delve more deeply into background details. This lays a healthy groundwork for setting up the sharing of more specific ways that efforts contribute to a field or body of knowledge.

Methods

Methods describe the steps that a team takes to conduct a study or run an experiment. The depiction of methods can be cumbersome, boring, and overwhelming. Weigh how much information you need to share to reach the end-goal of your film. Everyone has a limited carrying capacity for information, and pushing a viewer to go above this ceiling creates resistance against learning. Even worse, it may conjure negative associations to the content. So that you don't lose touch with our audience, return to the exercise of putting yourself into the audience's shoes. Determine the knowledge-burden that your intended audience can handle. Only the most potent storytelling, not the amount of information, can nudge a viewer to step above their comfort zone.

If your mission is to share the relevance of research or how it fits into a grander scheme of science or technology, the precise methods may deserve less attention than the background and later, findings. Return to the notion of audience's carrying capacity. Audiences who need large amounts of background knowledge will likely have limitations to assimilating the details of complex methods. This might mean that you either keep the methods section small or move through methods carefully, with clear references back to knowledge gained in the background section.

Results/Findings

You have now traversed a lot of material in your film, and have set your viewer up to grasp your findings. Congratulations, getting this far as it is not easy! As you formulate a storyboard that brings you to this point, keep this in mind: the presentation of your findings should align with the knowledge that you gave your viewers in the previous sections. This is a time to let the previous sections work for you. Rich background and unique methodologies will present impetus to digest the results.

Portrayal of technical findings is a craft. Graphs and tables might have been the medium that brought you, a specialist, to best grapple with your results. However these might not be ideal mediums for explanation to those less connected to the content. Technical fields have cultivated methods of communication through tables and graphs over hundreds of years. These techniques require a certain level of familiarity. Try to keep depictions of results modest or with means familiar to your viewers. If you are going to work with complex tools, gradually bring viewers into the light of how they work and why they may be advantageous.

Conclusion

You are over the hump. Hopefully your audience already gets what you are going to share in your conclusion. Conclusions are the place to coalesce and condense the important remarks from earlier in the film. In order to ensure that you end on a strong note, share where your body of work fits into the grander scheme. If it is curious, additive, or groundbreaking, say it. If it is going to instigate new efforts for refining particular methods in a field, use that as bait for people to stay tuned. The story does not have to end with your current findings, rather it can suggest why your efforts are relevant to building a new future. You have shared a great story so far. Now leave them with the most tantalizing morsel. Some people find the tail end of a film to be the time to leave a cliffhanger or a parting thought. This could be a great time for you to share why your follow-up research will be even more important. Leave them wanting more.

Final Comments on Treating a Film Like a Publication

I just described a simple prescription for how you can share a scientific story. Even if you choose to not use this method, the mind-set helps. The prescription emphasizes the value of information shared in a progressive and logical manner. It initiates discussion about the benefits to developing different acts of your story that account for the knowledge and level of interest of your audience.

Effective language maximizes how deep into details you can go and simplifies storytelling, even with technical content. Films are often less formal than research articles – the voice that you use to communicate thoughts and concepts benefits from a conversational tone and pacing. Even if you are catering a film towards peers who are experts on a technical topic and enjoy your written manuscripts, a little informality (in comparison to the voice in a research paper) lends approachability. If you are tempted to literally cut and paste text out of a paper into, say, a narration let this be the start. Then start rewriting the wording into a more relaxed script.

Storyboarding: A Dedicated Plan for Producing and Editing Films

Thus far, we've asked; what is the mission of the film? Who do you want it to reach? What is the style of your film? Hopefully these questions have started you down the road of determining what content you will include, and maybe how you will utilize various media (animations, interviews, music, landscape shots, narrations, etc.) to convey concepts and carry your story. And now, you know some basic rules for effective storytelling.

Let's now develop a more honed vision of how you will assemble your film(s). The next step is to build a *storyboard*. A storyboard is a blueprint that will describe how you will merge and sequence different forms of media to tell your story. A typical storyboard includes words from a script and chronological series of sketches or photos that represent the final media that you hope to use. Additionally, notes describe the tone of music and the words that are spoken. The images are accompanied by written word detailing how each scene is envisioned. All queues are meant to give you, the moviemaker, a clear vision of how each scene will look, sound, and feel. If storytelling is the foundation of filmmaking, storyboarding is a filmmaker's right hand.

While storyboarding, feel free to get excited with the story that you wish to tell. Enter this process with levity, intention, and focus. Blend your creativity and linear thoughts. I suggest that you pick places where you feel like you can freely express yourself. This is the time when you unleash all of your ideas and then massage them into a storyline that you (and your partners) believe will work. If you are a part of a team, grab your coffee and meet in a conference room. This is when you can share the kernel of a thought that ignited this moviemaking venture. Then pass the baton to the next person to help grow the vision. Be open to everything (you can always shoot ideas down as you get closer to a concrete vision). Never hold back on acting out what you are envisioning, and then jot notes down so that a week or year later, when you want to harken back to these creative moments, you will have a clear concept of what you had in mind. Jot ideas on Post-its, napkins, or even film your enactment of what will happen with your smartphone.

The more structure and details that you or your team can provide protect you from getting stuck spinning your wheels with time-intensive activities, which are later dumped. A well-constructed storyboard gives you the freedom to dive much more deeply into filming and editing, with a clear picture of each step's parameterization (e.g., specific media, characters, or locations). Once embedded in filming or editing, many people find it difficult to extract themselves from these activities to look at how one small step fits into the big picture of the story. With a storyboard,

Image 3.1 Many different types of media are incorporated into a film for success

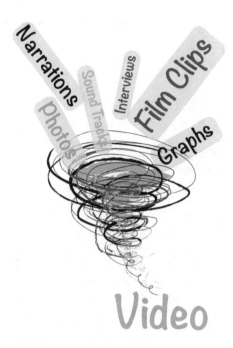

you are reassured that all of your efforts fit neatly into the bigger picture of your film. Don't sweat it. You have concise notes to guide you! (Image 3.1).

Some filmmakers, with whom you might collaborate, have a very formal framework for how they want storyboards to look. However, I have found this to be a rarity, and more loosely structured storyboarding works in most applications. I would say that until you are near a final storyboard product, use all the creative and exploratory methods that you can. Later, these ideas can be assimilated into something that fits collaborators' needs. Here I share a very general arrangement for storyboarding that likely isn't far from what others are using. Feel free to research or modify what I describe to suite your own needs.

The easiest way to start the storyboard process is to grab a handful of paper. Orient the paper as portrait (paper is taller than it is wide), and then scrawl a series of boxes just like you see below – six large boxes in all, three tall and two wide. I then segment out the boxes such that a small fraction, at the base, has room for notes. Each box represents a scene in your film. The farther down the page you go, scenes progress further through time. How much time each box represents is up to you. For a fast-paced film, the box might embody a couple seconds. For a much slower interview, a box might represent 45 s (mind you, 45 s of one scene may seem like an eternity through the eyes of a viewer) (Image 3.2).

Begin each scene with a title and a quick note on how long the scene will last. I put the title above or below each row of boxes. I also try to include the approximate amount of time that the scene embodied in one box will occupy in the film. For example, "Kayaking the Amazon River (0:20)."

Image 3.2 One storyboarding template

What goes into the larger boxes? They will be the home of imagery clarifying what each scene looks like. These can be very simple sketches or even photographs. There are innumerable images that you can grab off the Internet as well. Some cartoons can be found on websites dedicated to helping people with storyboards. Other images can be found from *Google* searches. The role of these images is to give enough information for the videographer and the editor (perhaps you will fulfill both of these roles) to have a clear understanding of the vision for each scene. Unless you publish your notes, all media is open for this stage or the moviemaking process. If all members know that the scenes will be from a specific video shoot you can keep the drawings very simple. If you are going to work with people who are outside of the storyboarding process, details describing the appearance of a scene will save you time and resources.

How do you include a script of what characters and narrations will say? Some filmmakers like to account for what characters are saying by putting word bubbles emerging from a person's mouth in the sketch. Alternatively, a script can be transcribed in the second, smaller box below the picture. This space is also dedicated to describing how the scene will progress. Write down the clear notes for on how you hope the scene will unfold. Consider answering how you will open the scene. Fading in from black? Describe, say, the increasing volume of drums? Use this space to remind yourself of the mood that certain music will elicit and how the visual elements within the scene progress. "The viewer will languish in beautiful shots of a redwood forest, with one shot fading into another. Gradually, the shots will progress from far away to within the forested thicket where we see birds nesting." Or, "The video will transition from a person knocking on a wooden table to a very simple animation of sound waves translating more rapidly through wood than air." You might be noting that this is a heap of information to include in a small space. If you run into space limitations with six boxes per page, fill a the page with only one or two boxes.

There are alternatives for building storyboards. A search on the web for "storyboarding" results in numerous links to simple and useful applications dedicated to easy storyboarding, eliminating all roadblocks to conveying creativity coupled with concrete visuals. A lot of them provide resources that you can print or manipulate digitally. You can drag and drop digital images from folders. Others have standardized cartoons to fill in for needed characters and settings.

If you have a mix between digital photos and hand drawings, print out the pages after the digital images are embedded. This way, hand drawings and notes can be incorporated. Unfortunately, as a storyboard matures, it is hard to create interactive edits on a printout. I have found that a more functional solution for incorporating hand drawings into these digital formats is to draw on a piece of paper and snap a shot with the camera on your phone (Image 3.3).

What I prescribed above is just one way to progress, keeping in mind that management of your storyboards is meant as a tool for YOU. Use what works.

Supplemental Storytelling Tools

A few filmmaking tools are worth contemplating before jumping headlong into writing a detailed script, complete with stage directions, filming directions, and editing notes. These are not guidelines for telling stories as much as gentle nudges towards approaches that are useful when transitioning your story into a potent storyline.

Image 3.3 The images that you incorporate into a storyboard can come from anywhere. Sketches, photographs, or stick figures all might do the trick

Storytelling Tools #1: Treatments: A Formalization of Your Plan of Attack

Treatments are great tools for organizing thoughts and intent for a film or series of films. So what is a treatment? They are hybrid documents, halfway between a mission statement and a full-blown script or complete storyboard. For a short film it may be one page, while others are well over 20 pages. Your ability to communicate your plan could serve as a sales pitch to, say, television producers; however it is also a very valuable activity for organizing your workflow, defining your audience, and unlocking what makes your videos unique, along with a healthy dose of why this film will be a smash!

A potent treatment paints a clear picture, with prose and thorough illustrative imagery of what you hope to accomplish with your film(s). They describe the methods that you will use to transport viewers from the beginning of your film, through a sequence of acts that build a plot and develop a crux, to a climactic resolution.

When writing a treatment, try to answer the following questions:

Why are you making this film?
Who is the intended audience?
What is your directing style?
Where will footage be procured?
How will you meet timelines and budgets?
Most importantly, how will you use distinctive tools to reach your decided audience?

Storytelling Tools #2: Watching Films Is a Multisensory Experience

Our senses have been honed for the survival of our species. Our ears, nose, mouth, touch, taste, emotions, memory, well-being (e.g., a feeling of balance), and that tingle on the back of your neck are tools for evaluating situations for threats or opportunity. Even when you are winding down to fall asleep, the natural state ready to be switched to on, ready to react. Indeed, most people use visual information to understand how to move through space or figure out what is going on around them; however, it is well known that sound is as important in making or breaking a film as visuals. Smell is the most powerful sense to connect people to old memories. Feelings and intuition are dominant over logic in decision-making. If you get nervous, there are physical feedbacks on, say, tightness in your chest or dryness of your mouth.

Potent films conjure experiences that engage a multitude of a viewer's senses. How does this work when most films only provide visual and auditory sensations? Viewers engage other faculties indirectly. They do this by associating events that unfold in the film with a viewer's own experiences (that might have been imbued with more dimensionality). Here are a few examples to ground you in what I mean. Ponder how you can use similar methods to add impact to your work.

You might have experienced these:

- A cold landscape, snow blowing, and the only sounds that you hear are wind and snow sifting through branches. The scene, and the ambient sound, might last an uncomfortably long time. Even a narration would detract from this feel. Scenes like this wordlessly force a viewer into feelings about being alone, vulnerable, and cold. This creates a profoundly emotional anchor for whatever content comes next. This is impact! (Image 3.4)
- A quagmire of traffic snaking through a cityscape. The movement of vehicles is sped up by a factor of 10. Cars lurch spasmodically through traffic lights like a lurching conveyor belt. Perhaps in the background you hear sounds of horns and engines. People rarely report that scenes such as these are relaxing. More accurately, people feel anxious, claustrophobic, or rushed. Blood pressure rises. This is the time when a filmmaker can dump a crux on the audience, say, a documentarian citing that people's happiness is decreasing as life's complexities rise. This is impact!
- A series of close-up shots on eyes, fingers hammering on a computer keyboard, and then computer code on a monitor switching to graphics (Image 3.5). The transitions between the shots are rapid. A computer programmer nods her head in understanding while grabbing a slice of pizza without so much as a glance. She pushes her glasses further up her nose and leans in eagerly towards the monitor. In the background you hear music much like deep space in *Star Wars*. We all know that the programmer is on a mission – motivated and engaged. She is resolving a problem, a big one, and might just figure it out. What will be the

Image 3.4 Successful storytelling recruits many more senses than the sound and visuals that define a film. Association to memories and reactions encoded into our memory and DNA makes for greater potential of lasting impact

Image 3.5 Films can tap into people's personal experiences, there by giving the storyline heightened impact

result of her work? All of us love when we are on a mission and singularly focused. We can practically smell the pizza from the last time that we did something similar. Do you feel a flutter of excitement in your chest? Perhaps this is when a third-person voice talks about there being no time to waste as an entrepreneur. Their product will simplify your life and allow you to focus on your dreams. You, the viewer, take your dreams and overlay them into the video, thereby giving the video personal relevance. This is impact!

Storytelling Tools #3: Don't Let the Quality of Your Equipment Hold You Back

Successful filmmaking does not take an immense amount of money. Depending upon your applications, there is a chance that you already have all of the necessary resources at your fingertips. In fact, you CAN build a storyline, edit, and film entire movies on many smartphones. Really! Modern smartphones are small computers that manage countless applications, and only one application makes calls. The image sensors for the onboard cameras of many phones produce very high-quality, high-resolution videos. With some, you can even film slow motion or time lapses. Mind blowing and opportunity creating!

More importantly, the greatest tool that you have is your mind. Leverage the power plant between your ears (your brain) to figure out all of the resources at your fingertips. With a bit of strategy and creativity, you can turn annoyingly poor quality footage into endearing or catchy films. Perhaps this means that you tap into a viewer's nostalgia by overlaying an effect (during editing) that makes a poorly captured video look like it was recorded on an old, grainy film camera. This takes their mind off of the poor composition to enjoy a walk down memory lane. What if you use crayons for drawing and return viewers to experiences of being a kid? Draw an oversimplified sketch of a device, while covering up your lacking skills in rendering hi-tech infographics. Little does the viewer know that you are strapped for methods. Before they catch on, their minds are entranced by the notion that learning about the device is simple and fun.

Storytelling Tools #4: Curiosities Hook Your Audience into Caring

Some films are engaging because they present a chance to learn. The audience's curiosity about fascinating, engaging, or relevant topics is the driving force for watching. Such videos ask people to ponder concepts because their own interests, experiences, or professional goals override the ability to turn away. If this is the case, we, the filmmakers, have to know what questions are going to lure them in.

Step back to your mission statement and ponder for whom your film will be made and what will draw them into your content. You have had months, years, or maybe decades to build an understanding of your field of work, form your own opinions, and foster dedication and allegiance to relevant information. The same is not likely true for many other folks, so carry them to realization through queries that align with their level of interest and experience – in other words, within their comfort zone. More importantly, since you have some level of authority for the material, you can strategically draw upon the tastiest morsels to hook their attention.

What could this look like? If you are interested in sharing the value of solar cells for making electric energy, perhaps start them off with a question, posed by a narrator or host, that most people can answer: "What is your favorite appliance?" As you progress through your film, your questions can involve more sophisticated thinking, knowing that the foundational knowledge that you have conveyed will lend adhesive forces to new knowledge. Let's say you wish to discuss the role of photons for powering solar cells. Prime the viewer with how photons relate to their lives. "What is it about the sun that travels through massless space and warms your body on a sunny day? Photons!"

Return to empathy. Try to ask the questions that hooked you into interest. If data suggest a crucial change in a condition, simply ask the question that came to your mind when you first saw it. This can be conveyed through "how, when, where, what, and why" questions. A personal example – "For hundreds of thousands of years, Earth's temperature tracked the amount of the sun's energy hitting the planet. In the 1980s that trend shattered. Why?" By voicing the questions that you address in your work, you are showing that your work is driven by curiosities. You are keying the viewer in on the adventure of being a scientist, not just being a body of information. (By the way, we believe that the answer to the question above is because of the increasing concentrations of the greenhouse gas CO_2 in the atmosphere.)

Asking questions of the viewer is a delicate art. When done well, a few questions can facilitate personal connections between the audience and the content. However, too many questions can be overwhelming. Additionally, audiences are sensitive to loaded questions. They sniff them out. If you want a person to think a certain way, questions may seem leading, off-putting, even passive-aggressive. Avoid asking such questions as, "Do you like it when you lose money?" No one likes to lose money! Use a more personally engaging question transporting them to the same end: "How do you feel when the stock exchange plummets?" Yes, you are steering a conversation, but asking them to make valuable connections through their own sensibilities and emotions.

Story telling Tools #5: The Seven-Second Hook

Think about the startings of films like meeting a new person. We get a sense for who someone is in the first few seconds of interaction. The tone, music, steadiness of footage, background noise, and many other attributes amount to a person's first reaction to a film. For many, this rapport resonates for the remainder of the film. Try a simple experiment. Explore videos on *YouTube*. People often cite that splashy graphics introducing a film is far better than jumping headlong into the body of the film. Oppositely, you may find that filmmakers positioned the title of a film half a minute after the storytelling starts. An early title might not be efficient at creating interest in the film out of the gates. This is a healthy nudge to think about ways to knock the first handful of seconds of your film out of the park! Within film, this is called the "Seven-Second Hook."

Final Thoughts on Storytelling

Storytelling is a skill. Although there are no rules for how great stories must be told, there are suggested elements and ways to develop story arcs for maximum impact. Every film tells a story, and it is up to you to increase the chances that resulting impacts on an audience are what you hoped for. Time and practice will make good storytelling feel second nature. Combine this with a advanced research and planning and your are upping the chances that the end results will be fantastic.

Chapter 4
What You Need to Know About Video Files

What You Need to Know About Video Files

Awareness of frame rate, resolution, and film formatting is important for understanding why some video files are larger than others. These characteristics of film are large determinants for how we make and distribute our films. For example, the Internet may be a great platform for the exhibition of your film; however, many homes and offices around the planet are lacking Internet speeds adequate to transfer the vast amounts of data that come with high-resolution videos. Decreasing video resolution might better suite purposes to reach people in developing countries. Fortunately, technology is changing, so this dynamic is shifting by the minute.

Similarly, data volume challenges easy use of computer and film electronics. In order to wield large/high-resolution video files, computers must come equipped with vast hard drives for video archiving, large random access memory (RAM), and fast video cards. The higher-end cameras and computers that handle these burdens come at costs that can strain a production budget. If such devices are not feasible for you, editing with lower-performance computers are slow and crash routinely under load. With both web and equipment limitations in mind, filming with the highest resolution cameras that you can buy may not be the smartest option for you.

I urge film makers to work with the best quality footage that they can wield. Larger video clips might be cumbersome from a file storage (size) and manipulation perspective, yet they make for undeniably great viewing. Even more importantly, once a scene is recorded, it is impossible to improve the film resolution above that of the original. What's more, low-resolution or low-quality clips limit its usage down the road. Your work might have the potential to explode on a scene and yet its resolution makes it less attractive to television networks. Additionally, returning to rapidly growing computing capabilities and web speeds, in a few years you may be kicking yourself because your great film cannot be exported in a resolution for your uses.

R. Vachon, *Science Videos*, https://doi.org/10.1007/978-3-319-69512-9_4

An appreciation for frame rate, resolution, and film formats will help you navigate many of the future discussions in this book, the first of which will be the purchasing of your filming equipment.

Frame Rate

Movie scenes show movement because video strings numerous sequential still shots together, end to end, and play them in rapid succession. Images change so quickly that the little differences between each shot manifest in a seamlessly unraveling scene. A camera's *frame rate* is defined by how many frames it captures during every second of recording. The more frames per second, or the higher the frame rate, the smoother the video. However, with higher frame rates (at a given resolution), the more information that each video file occupies. What's more, there is diminishing return on smoothness beyond certain speeds. Most video cameras record at either 23.976 (which is frequently called "24 fps") or 29.97 frames per second (simply called "30 fps"). In general, the 24 frames per second speed is used in movie theater expositions, while the 30 fps align with the frequencies of data transfer on television and over the Internet. The film industry is currently exploring ways to use film at 48 fps. The differential frame rates highlight the importance of knowing where you hope to disseminate your videos before you start filming. That way, you can invest in a camera with the proper settings and later create a film at the appropriate frame rate.

How many frames per second make for a smooth video? Early in the days of movies, films were projected at between 16 and 24 fps. Today, the standards (24 and 30 fps) work fantastically for almost all situations. However, some applications ruin the smooth playing of scenes captured at these standard film rates. For example, the making of slow-motion film-clips stretches a scene out to show action more slowly than it originally happened. Someone ski jumping may take on an ethereal and dramatic feel in slow-motion. Let's say that you slow down a 1 s clip to last 3 s. If you filmed at 30 fps, the slowing down of footage means that the 30 frames from the original second of footage are stretched to 10 fps over 3 s. This then begs the question: will the footage still play in a smooth manner?

The utility of slowed down clips depends upon what you filmed. If we continue with the ski jumper example (with large, obvious moving elements) a slow-motion clip will likely appear uncomfortably stilted. In such situations, I avoid slowing down scenes filmed at standard film rates, unless absolutely necessary. Differently, if you want to slow down a scene with very few moving elements, such as a faraway cityscape, I believe that you can readily reduce the number of frames per second to half speed, so perhaps 12–15 fps. As such, the answer to the question about what video frame rates work best lies primarily in evaluating the content that you are recording, the context of its use, and simply how it looks.

Many modern cameras give you a much better alternative for slow-motion filming – pushing the recording rate to higher than 24 or 30 fps. For cameras that come with such an option, frame rates typically increase by even factors, such as times two or four. This means that in some camera's menu, you can increase your frame rate from 30 fps to 60 or 120 fps. If you film in 120 fps, you can produce slow-motion footage at quarter speed while maintaining a frame rate of 30 fps. If slow motion is something that you think might add value to your films, look into purchasing cameras with higher frame rate options.

Note #1
Filming on high frame rate settings, over long periods of time, will rapidly fill up video cards and burden computer processors. Since these burdens also challenge the processors within cameras, higher frame rates may only be possible with associated reductions in resolution of the recordings. These reductions in resolution may be more deleterious to your final film product compared to the benefits gained from a slow-motion scene. These detractions are for you to weigh.

Note #2
Bitrate refers to how quickly data is transferred from one place to another. Some people link bitrate to pluses and minuses of film, yet, more accurately, it has to do with the strengths and weaknesses of the devices that are recording, transferring, or broadcasting the footage. Indeed, larger video files are harder to manage if bitrates are slow, yet discussions about bitrate are only tangentially connected to the qualities of footage.

Resolution

Resolution refers to the number of pixels that each frame contains. Pixels are the square dots within a grid that fills a shot. In order for an image to manifest as something recognizable, your camera defines each dot by a color. When the colors of all the pixels within a scene are seen in entirety, your image will resolve into a landscape or person's face (Image 4.1).

The measure of the width compared to the height within a shot is the aspect ratio. Video aspect ratios are typically 16:9 or 4:3, meaning that videos are wider than they are tall. In fact, for the past decade, most modern video cameras record in 16:9 and very few in 4:3. In most recent years, 1.9:1 is an aspect ratio used by the very highest resolution cameras. Every frame that a camera captures not only retains the aspect ratio but holds constant the number of pixels in the horizontal and vertical dimensions. They can be later stretched and deformed in editing programs. Unfortunately, increasing the size of a video after it has been filmed will not

Image 4.1 When one zooms in on a digital image, one can see the individual pixels that constitute shapes and colors. Together, the pixels configure to form an image

increase the visual clarity of a shot. Quite the opposite – altering video footage resolution, even to fit a larger video resolution, degrades footage quality, at least slightly.

The number of pixels in a scene is calculated the same way that you measures the area within a square – multiply the number of pixels going across a frame by the number of pixels in the vertical domain. If a clip has 10 pixels going across by 7 up,

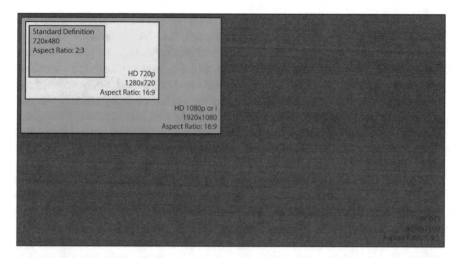

Image 4.2 Here we depict the aspect ratios and proportional comparison of standard digital recording sizes. Although higher resolution portends better footage, its large size can burden slow computer processors. Additionally, high resolution, such as 4K, may prove to be overkill when the place that you hope to screen your films only displays movies at a lower resolution, like 1080p

the clip will have 70 pixel resolution. Indeed, if a video clip really had 70 pixels in total, your scene would have incredibly low resolution. As such, this number is dysfunctional for filming scenarios; however, the smaller numbers make multiplication in this example easy.

Higher-resolution shots allow the viewer to see greater detail in a scene. For quite some time, televisions were limited to 72 pixels per inch (high-quality photos and prints are typically printed at more like 300 pixels per inch). As such, a full screen that is 20 inches wide would show 1440 pixels across. With such a case, a video with a width greater than 1440 pixels would be overkill, because the screen resolution is the limitation to your viewing experience. Although most modern televisions and monitors have resolutions much greater than what we discussed, the numbers that I mention are used to help you wrap your brain around how screen pixels relate to the resolution of footage.

Approximately a decade ago, digital video capture started to migrate away from tapes towards solid-state memory cards. Around the same time, high-definition video became increasingly more common. Most filmmakers would describe high-definition video as any number of pixels greater than 720 × 480 pixels (the standard resolution of film a couple of decades ago). Today there are two basic pixel measures that fit into the high-definition category: 1280 × 720 and 1920 × 1080. If you do the math, these numbers result in aspect ratios of 16:9. Some people simply say that films are "720" or "1080" when referring those resolutions mentioned above, respectively (Image 4.2).

Oftentimes, you will see 1080 resolution written as either "1080i" or "1080p." Both video formats have dimensions that are the same; however, "i" refers to *interlaced* and "p" refers to *progressive*. Interlacing is a system that was invented to increase the perceived resolution of a video while not making associated video files unwieldily large. Interlaced videos require special recording or broadcast electronics that alternate displaying even and then odd numbered rows of pixels within a video clip. If you see odd numbered rows in one clip, you will see even numbered rows in the next. An image of odd or even rows is called a field. So, in order to see a full image, one field is rapidly followed by the next. How many rows do you see in one instance in a video that has 1080i? 540. To the eye, scenes appear sharpened; however, scenes may flicker a wee bit. Differently, a progressive scan video shows all rows in frame. At the cost of larger file sizes, the video quality of "p" over "i" is noticeably better.

Most cameras record with resolution of 1080p, but some modern, and often expensive, cameras also record in "4K" resolution. This resolution means that approximately 4000 pixels are captured in the horizontal domain. The precise number of pixels varies between two different standards. Ultra High Definition (UHD), the lower resolution format of the two, has pixel measurements of 3840 × 2160. This maintains a 16:9 aspect ratio. Perhaps the more universally accepted format in cinema is Digital Cinema Initiatives (DCI). DCI changes the aspect ratio to 1.9 × 1 and a resolution of 4096 × 2160.

If you are looking at televisions or video cameras, companies may pitch their devices as 4K, when they are referring to the UHD format. Many folks in film think it more appropriate for "4K" to refer to DCI and frown upon it also referring to UHD.

Regardless of the precise resolution that 4K refers to, it contains approximately four times the number of pixels as high-definition (1080) video. Your computer could struggle as a result of the increased data associated with 4K video. Perhaps someday soon 4K video will not challenge the computing capabilities of everyday laptops, but as for right now, check to make sure that your system will accomplish all the functions that you need under these stresses before you purchase a camera that films in 4K.

Compression

When jumping into videography, it behooves you to learn about how video files are recorded and stored. First, when you open movie files on your computer, they have extensions (the last letters of a file name following a "."), such as .MOV, .FLV or .WMV. The extensions define the formatting for the particular video that you watch. What does *formatting* refer to? Formats create *container* files that house video, audio, subtitles, and more media together. The three examples mentioned above refer to Apple's *Quicktime*, *Flash Video*, and *Windows Media Video* formatting, respectively. By any one of these, different allgorithms weave all files in the container into one complete movie experience.

Most video and audio files are compressed to one degree or another. Compression is the process of cameras or computers reducing the amount of data that files occupy.

Normally this begins with electronics recognizing patterns in visual or audio information. Consider this – some image compressions identify regions or shapes within a scene that are all one color, rather than the totality of individual data points, pixels, within a shot. By defining colors of a shape, rather than by every unique pixel, file size is reduced.

Unfortunately, some compressions will lower your video quality in noticeable ways. For example, if you were to film a scene in which the sun is creating a halo in slightly veiling clouds, the varying rings of brightness around the sun may take on a look of concentric colored donuts rather than a very natural and even transition from light to dark. Because of compression algorithms simplifying the data within a scene, working with compressed files means that you are working with files that are lacking in quality. As such, while working with a compressed file, you might later find that a clip is disappointing or inadequate. Reduce your risk to these let-downs and use original footage whenever possible. If you hope to use other people's footage, make special requests for uncompressed files.

Part and parcel of playing a video file that has been compressed (along with sound and subtitles) is encoding and decoding. The compressed files need to be translated into media that you can see and hear. The programs that encode and decode these compressions are called *codecs*. Fortunately, your computer comes with numerous codecs for playing movies that have been compressed with different programs. Should your computer tell you that a video cannot be played, scour the web for video players compatible with it.

I heard someone say that the *container* is the box that holds a video and the *codec* is the crowbar that forces the video to fit the right size box. There are several *containers* and *codecs* associated with different computer operating systems (like macOS) and more functional under certain circumstances. Here I share information that might seem overwhelming to some; however, the knowledge is helpful as you choose how you will film and export your final movie.

Popular Containers

AVI and ASF: For a long time, Audio Video Interleave (AVI) was Microsoft's primary container for movies recorded and played on their operating system. AVI's popularity has dwindled in part because it is not as compressible as other formats. Advanced Systems Format (ASF) is another Microsoft container that leans on Microsoft's Windows Media Video (WMV) and codecs. These files are typically suffixed by the codec name, rather than "ASF".

MOV or QuickTime is Apple's container for films. They have grown quite popular because of their compressibility, compatibility with numerous codecs, and options for changing their aspect ratios.

Advanced Video Coding, High Definition (AVCHD) is a container developed by a joint collaboration between Panasonic and Sony. AVCHD supports standard and high-definition videos (and newer versions support 3D).

Flash Video (FLV or SWF) is a long-standing container that is known for easy streaming over the Internet. While quite functional, it is not supported on many Apple-based devices.

WebM and WebVTT are video formats designed with rich understandings of modern Internet technologies in mind. WebM was developed as an open source platform while WebVTT was developed by the World Wide Web Consortium.

Popular Codecs

H.264 is a very versatile codec well known for managing standard and high-definition digital videos at high and low speeds (bitrates). This codec is very popular with digital video cameras.

Windows Media Video (WMV) codec is often used with high-definition video and audio streaming in that they allow the film to start playing before it is completely downloaded, thus saving the viewer time.

Popular Formats That Are Both Codecs and Containers!

MPEG (1 and 2) are a couple of the more used formats for use with DVDs and Video Compact Disks.

MPEG 4 (also called MP4) format contains video and audio tracks that are compressed separately. This format is often used for web applications.

(References: imagenevp.com and videomaker.com).

Final Thoughts on Video Files

There are a lot of details to understand about video files; however, there are a couple of qualities that must be met for you to move forward. First, make certain that your video files are of adequate resolution to fulfill the requirements of your project. Second, video files can be enormously taxing to camera and computer processors. What's more, large files can be very challenging to transfer when Internet speeds are lacking. Take care to not overburden the systems that you depend upon. Otherwise, your efforts for quality could stymie progress.

Chapter 5
Purchasing Production Equipment: Cameras

Personal Anecdote

I got into filmmaking because I really liked the idea of making a movie. I was drawn to telling exciting visual stories, and I had no idea how easy (or hard) it was to make movies. At first, my attention was focused on producing movies that would grab my friends' attentions. Trips into the wild were a common theme. My transition to professional filmmaking was serendipitous, but it also hinged on some wise decisions I made – that at the time I never knew would result in success on international television.

I got my foot in the door of network television when I was an intermediate videographer. I had quite a bit of experience with photography, yet I had zero instruction with film. The first step was to purchase a very nice camera. This camera (a Canon GL2) was the lowest priced documentary-quality camera on the market at the time, but still recognized as a great workhorse. Indeed, investing into an expensive (for me) camera left me anxious because I knew that I could continue to make fun films using cheap cameras. Nevertheless, I wanted to see what I could produce with a better camera.

The summer I purchased the GL2, I brought it on a research campaign in the Andes Mountains. Unexpectedly, the footage that I captured supported some very fascinating discoveries. A couple very high-level research papers blossomed around the science, and network television wanted to tell the story. No one had footage like mine, so producers from PBS, National Geographic, and the Discovery Channel knocked on my door! The combination of being at the right place at the right time, and the desire to film and capture high-quality footage landed my clips in over a dozen news stories and documentaries.

What was my take-home message from this experience? (Image 5.1).

Pushing my budget as far as I could created opportunity. Quality equipment increased my chances to take my productions further than I expected. What's more, I learned that I should take more time, exercise more patience, and press record more often. Great footage gives more choices for how I tell my stories. I am guessing that my experience is not out of the ordinary, so I would advise fledgling videographers to push the quality of their gear as far as they can without breaking the bank.

© Springer International Publishing AG, part of Springer Nature 2018
R. Vachon, *Science Videos*, https://doi.org/10.1007/978-3-319-69512-9_5

Image 5.1 High camp in the Peruvian Andes

Purchasing What You Need

Filmmaking starts with an idea, but at some point you have to turn your attention to purchasing specific equipment and software to complete the tasks. This is an exhilarating time, and best executed with smart purchases. Certain gear is better for some uses over others. While one purchase might dovetail nicely with certain other gear, another might not. For example, some cameras couple with auxiliary microphones (a microphone that is not built into the camera). Some do not. As such, some cameras allow for greater growth into more complicated productions. There are also niche items, like unmanned aircraft systems (UASs – commonly called drones), that produce very specific and compelling outcomes.

From this point on, let's refer to the equipment and software that you will need as a *quiver*. This is not a term universally used in filming, but I hope that it will simplify some of our discussions.

The quiver that you end up using is often constrained by a budget, which in turn may restrict the outlets that you use for showing or disseminating your film. In an attempt to help you with your quiver procuring needs, I will break equipment and tools down into groups and discuss their pros and cons. When discussing tools for your quiver, I see four primary groups that you may fall under. These are loose boundaries for buyers. They are generalizations but let you know that no matter how flush or meager your budget, you have great options. Additionally, it pays to be informed because poor investments loom.

Group 1

- You want to try to make a movie and see what happens.
- Budget – less than $500.
- Viewing Platform: YouTube-like web outlets.

Group 2

- You are dedicated to making films of integrity and you want options to grow their applications.
- Budget between $500 and $5000.
- Viewing Platform: Web-outlets to film festivals.

Group 3

- You might be part of a team that hopes to make a body of work that will turn heads and have the potential for many applications.
- Budget $5000–$50,000.
- Viewing Platform: Broadcast-quality outlets. You don't want technology to hold you back from conveying your concepts through film.

Group 4

- You want to create high-level film productions bound for network television and feature-length Hollywood movies.
- If you are imminently ready to jump into work of this scale, this book alone is likely not ideal for your needs. You will need considerable experience and partnerships to take on tasks of these magnitudes.

In all groups, purchases must be well thought out and executed. For example, If you are in Group 2 yet can't imagine making the film that you have storyboarded because it is too expensive, read up on all gear options. The hope is that some of the less expensive gear or creative solutions can get you over some intimidating humps.

Money won't buy you a good-quality film. If you are not interested in investing a sizeable chunk of your schedule into learning and practicing, the films that you produce will have limits. There is no substitute for time in the saddle. However, if you have a sizeable budget for gear and minimal time to invest, I would suggest focusing your searches on the higher-end gear that has very functional automatic settings. As an alternative, several moderately priced devices provide user-friendly interfaces and produce great results. A consequence of investing into moderately priced gear is the opening of funds for hiring an experienced videographer, taking a weekend videography course, or purchases later in the production process (like animations), which might lend tractive qualities to your film.

Alternatively, perhaps filmmaking is just a side project for you. To get started, you might not need to purchase any equipment at all. Simple solutions can be found in equipment rentals or borrowing gear from friends. Rentals are a fantastic way for you to learn whether filming is something that you would like to continue doing. For a small investment, you can get firsthand experience different equipment. Additionally, you can start to use the industry's hottest gear without dropping huge amounts of money.

Camera Types

Purchasing camera gear values from understanding the different resolutions and formats of video footage (Chap. 4), Familiarity empowers you to more easily research the cameras that are compatible with your budget and needs. There are numerous types of cameras that are built for specific applications. With a limited budget, you could be shooting yourself in the foot by running out and purchasing a camera that gets incredible reviews but is best suited for uses other than yours.

Here we will break down the pros and cons of the smartphones, sports cameras, Digital Single Lens Reflex (DSLR) cameras, and camcorders.

Smartphones

Let's start from the lowest buying potential, Group 1. This could mean that you want to spend as little as possible on a filmmaking effort. Smartphones are great tools when your budget is tight. The image sensors and video quality of modern smartphones are mind-blowingly good for the price (Image 5.2). Indeed, many smartphones give you the option to film in 4K resolution and come equipped with sophisticated software that helps to stabilize your shots. Even better? You can purchase lenses that fix onto the outside of your phone to increase your options while filming. Simple, elegant, and free (if you already have a smartphone).

Despite their utility, smartphones have a few notable drawbacks. First, without auxilary lenses smartphone cameras zoom digitally. Unfortunately, video resolution is lost in the process. Digital zooming selects a subset of pixels within a high-resolution shot by cutting into the center of the image sensor, eliminating the region that surrounds this area. As a result, the final zoom is defined by fewer pixels which is a lower-resolution shot. Zooming with a lens is preferred.

Second, the image sensors on smartphones are not ideal for all lighting conditions. Since phones are so small, the image sensors for their onboard cameras are designed to be tiny. In general, small image sensors are great in well-lit applications where there are not huge contrasts between bright and shadowy areas. Small image sensors are not very good at showing detail in both bright and dark regions. What's more, faithful color reproduction becomes increasingly poor in darkening spaces.

Image 5.2 a. Smartphones are compact and portable computers, with many applications, including filming. **b**. Most modern smartphones come equipped with excellent built-in cameras. If budgetary limitations are dire, they can go a long way to meet your filming needs

Lastly, capturing good-quality sound with the onboard microphones is very challenging. Microphones on smartphones are not of the highest quality. What's more, smartphones give you very little latitude with capturing acceptable-quality sound when the sound source is far from the camera. Imagine trying to interview a person in a mall. Malls are always bustling with activity and background sound. In order to capture optimal sound from the interviewee, it would behoove you to have the microphone pretty close to their mouth. This is a challenge when you also might want to film from 8 feet away – in order to frame them within an image correctly. While we go into useful solutions to this problem in the chapter on sound capturing techniques (Chap. 13), it is important to know that you can purchase auxiliary microphones that couple with smartphones. These can connect with the phone either through the earphone jack or with Bluetooth technology.

Sports Cameras (E.g., GoPro)

Technology has bestowed the outdoor world with the ability to film some of the most amazing activities and environments with video. Tiny cameras are mounted to helmets, handheld sticks, or the tips of surfboards (Image 5.3). The utility of these cameras stems from sophisticated design and modern electronics. For ~$400, placing them barely within Group 1 and more comfortably into Group 2, you can film underwater or in slow motion. Even more incredibly, in very high resolution (4K in some cases)! I believe that the options to record sound from auxiliary devices that many sports cameras offer makes them indispensable. But there is more. Their depth of focus is quite large, meaning that they can focus on objects far away or practically up against the lens, sometimes at the same time. For some applications, this is a powerful asset.

Small sports cameras present enormous opportunities for filming; however, users should understand the animals that they are. Since these devices are designed to capture activity, they use a wide-angle lens. One outcome of a wide-angle lens is the a perceived reduction of shakes while recording. Whether fixed to your helmet or held by hand, movement will produce instability in your shots. Wide-angle shots minimize the impact of these bobbles in your resulting shots.

On the other hand, wide-angle shots also make nearby objects appear far away and often distorted (as though they are seen through a fish eye). What's more, far-away objects feel really far away. If you need any shots besides those that seem expansive, a sports camera's utility wanes. Fortunately, some of the more advanced versions of sports cameras (like the GoPro) give you the option to change the default settings so that a recorded scene appears less wide-angle. To do this, you program your camera to capture the center of the wide-angle shot, thereby reducing the wide-angle, expansive feel (Image 5.4).

Some of these cameras do not come with (or pay extra for) an LCD screen for viewing your shots, making it difficult to review your shots while filming. The more

Image 5.3 A sport camera in action, mounted on the back of a household pet

reputable companies like Sony or GoPro provide applications for controlling and viewing the camera through your smartphone. These are workable stopgaps to the problem; however, I have found that filing becomes tedious if you are taking numerous shots in short succession.

Should you have need to capture activity from afar, like lectures from the back of a classroom, sport cameras are not recommended. DSLRs and camcorders (discussed soon) excel at these uses. Interviews with a sports camera require some forethought and orchestration. If you are planning on filming in controlled environments, you might have practical options to position a sports camera up close.

If your limitations and goals place you into Group 2 or 3, sports cameras can deliver great results as a secondary camera. Use one camera for, say, conducting interviews, while your sports camera shows what it is like to dive on a ship wreck, sample gases from 40,000 feet into the atmosphere, or examine the habits of newly hatched falcons from right beside their nest.

Image 5.4 Sports cameras capture incredibly high-resolution videos and photos. Because of their small image sensors and high technology, one can film with both the foreground and background in focus. Because of its intended uses, the lens is wide-angle, almost to the degree that it can be called a "fish eye." If you wish to use a sports camera for applications that benefit from more mid-range angles, some cameras give you options to crop into the center of the scene where the curvature of the lens has less impact on the content within its framing. Here we cropped in on a sports camera photo to illustrate how this works

DSLRs and Mirrorless Cameras

Digital single lens reflex (DSLR) cameras allow the light from the scene in front of the camera to pass through the lens and off of a mirror that sits in front of the image sensor. The mirror reflects the image through the viewfinder (Image 5.5). This lets the user see the precise framing of what will be recorded onto the image sensor, once the mirror is actuated out of the way, (which happens during filming). Mirrorless cameras are very similar, except they don't use the mirror to produce the image that you review through your viewfinder. Instead, the preview is digitized.

I prefer to use DSLR and mirrorless cameras for filming in most situations because of the exceptionally high quality of their shots. The image sensors on these cameras were originally designed for still shots so they are larger than most other video cameras. Video clips resulting from these larger sensors have great detail and their presentation of colors is rich. What's more, the large image sensor gives you a great deal of freedom with your focal range. We will delve more deeply into focal range in the chapter on filming methodology; however, simply said, larger sensors allow you to keep the foreground in focus while the background is not (and vice versa). This creates potent cor compelling visual effects. In general, the image sensors on mirrorless cameras are not quite as large as DSLRs, but this is not a rule.

I also lean towards using these cameras because they give me the option to take incredible photographs. Other types of cameras also take photos, but few can match the impeccable quality of a produced by such large image sensors. If you would like to mix and match photography into your work, mirrorless and DSLRs are serious workhorses. In this vein, taking photographs during a video shoot can be very useful towards your end goals. High-resolution photographs can be zoomed in upon in far

Image 5.5 DSLR and mirrorless cameras come equipped with large image sensors. DSLRs use mirrors so that users can look out the lens and see exactly what the camera will film. This is illustrated as we look back through the lens and mirrors on a DSLR to see the light emanating from the window behind the viewfinder. While we are looking at light behind the camera, this shows the routing of light as it passes through the entire camera

greater detail than the video clips (maybe you want to show part of a device that you could not easily capture in video mode). It is true that the elements in the shot will not be moving, but later, during the editing process, you could add dynamism by zooming in on the photo. We will discuss this more soon.

Image 5.6 Mirrorless and DSLR's large image sensors excel at capturing color and details, making them ideal portraits and landscapes

Mirrorless and DSLRs excel at certain shots (animals, facial expressions, landscapes, and hi-tech analyzers, as opposed to, say, exciting action from a helmet) (Image 5.6). The base price for mirrorless and DSLRs, complete with a very functional lens, is about $500. Costs rise to several thousand dollars. In my view, prices for decent, new DSLRs or mirrorless begin around $700. This means that they fit nicely into purchasing Groups 2 and 3.

DSLR and Mirrorless Drawbacks

DSLRs and mirrorless are not without their drawbacks. Let's go over a number of them. However, know that with experience, many of the challenges can be overcome and in fact give you latitude for more creative shots.

- Some DSLR cameras may overheat in video mode. As mentioned before, DSLRs were originally designed for photography. The cameras are small and the electronics compact. In video mode, the cameras are working over time to condense the information from their large image sensors to a certain number of frames per second. Every single DSLR that I have owned has gotten warm with vigorous use and a few have become sluggish from excessive issue. So, if your camera suffers from excessive heating, make sure to give your camera time to recover between shots or have a back-up on hand should your primary camera need a break.
- DSLRs and mirrorless are designed differently than, say, a camcorder. They are less ergonomic for video uses, and thus may seem cumbersome and awkward to wield. It's possible to overcome some clumsiness with technique and stabilizing mechanisms. Unfortunately, the problem does not go away completely.

Image 5.7 CMOS technology does not capture a scene on an image sensor all at once. Instead, each frame is captured more like a rolling scanner, so rapidly moving objects in a scene may have strange appearances. Here, the propeller on a plane looks to be disconnected from the engine hub. Vestiges of this process can be observed in resulting film

- These cameras have large rolling shutters. Rolling sensors are the foundation of how the CMOS sensors work (the image sensor frequently used in video cameras). A rolling sensor does not record a scene on the image sensor all at once. Instead, if you are filming at, say, 1/30th of a second for each frame, within this fraction of a second, only parts of an image sensor are exposed to the scene at a time. By the end of the 1/30th of a second, the entire image sensor is covered. This style of video capture can manifest in distorted footage, namely when filming a rapidly moving object or when your camera is not completely stable. Indeed many camcorders have rolling shutters as well, but they are not as large. This problem is much worse in DSLRs and mirrorless than camcorders because of the high image sensor resolution (Image 5.7). The simplest way to work with movement created by an unstable camera is to take extra care to use a tripod or hefty image stabilizer.
- Recording times of DSLRs are usually limited to 4GB file sizes. For high-definition filming, this limits you to approximately 9–12 min of filming or less. Should you exceed the size limits of your camera, a new clip may be initiated or the recording might stop altogether. Most video applications don't require continual

recording for such long durations, so this quirk, is not a show-stopper. What's more high-end higher-end cameras work around this obstacle with muliple cards. Simply be aware of this looming constraint as you go shopping.

- DSLR and mirrorless cameras usually come with detachable lenses. They are expensive! This poses the question, why invest in great glass (a term used for lenses)? Dedicated lenses improve the quality of a shot with nearly faultless glass. These larger lenses also permit lots of light to hit the image sensor at varying focal lengths. This is not as important when filming shots with conservative focal lengths or decent light conditions. However, large lenses afford more flexibility in variable or suspect lighting conditions.

- Focusing and zooming with most DSLRs and mirrorless is complicated. You can autofocus a scene before you start filming; however, not all of these cameras autofocus while the camera is recording video. In most cases, after pressing record, your camera will maintain focus on an object. If objects are moving towards or away from your camera, there is a chance that they will move out of focus. You will later learn that you can adjust the aperture of your camera to increase the depth of focus and thus keep elements at different distances from the lens in focus. Another solution is to switch over to manual focus and manually turn the focus adjustment barrel to maintain focus on the moving object. Unless your camera is very well stabilized, this often results in some distracting camera movement. The same is true if you want to zoom in or zoom out while you are recording. With most DSLR and mirrorless cameras, you have to zoom by hand, which is cumbersome. Like with manual focusing while you record, zooming manually can cause troubling camera movement.

- Most DSLR cameras have audio jacks for which you can connect auxiliary microphones. This is fantastic! This gives you options to record great audio tracks as you film. However, many DSLRs don't have earphone jacks for reviewing clips or hearing the sound that your camera is capturing. This limitation has plagued sound recording, so more DSLR cameras are coming equipped with these ports. As such, if you are shopping for a DSLR and mirrorless cameras, keep your eyes peeled for both an in and out jack. There are other work-arounds for this issue, unfortunately, these are another piece of hardware that you might have to tote from place to place (Image 5.8).

Camcorders

Motivated, introductory videographers working with a well-designed camcorder stand a really good chance of capturing very useable footage (Image 5.9). Why? Camcorders are jacks-of-all-trades. They come complete with options to cover a large number of scenarios. They zoom smoothly, capture great sound, and monitor video and sound in real time. Likewise, their image quality is decent.

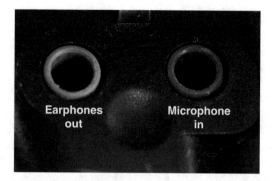

Image 5.8 DSLR cameras usually come equipped with an auxiliary microphone input jack; however, many do not come with an earphone jack. This means that you would not have an easy way to conduct real-time audio monitoring. If you shop for a DSLR with the intent to film video, try to find one with both microphone and earphone jacks

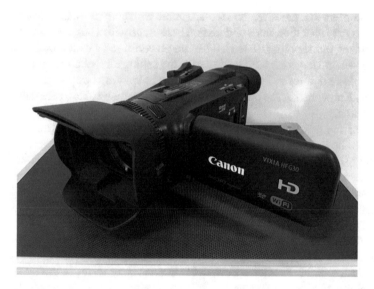

Image 5.9 Camcorders are dedicated to fulfilling as many of your video needs as possible. The accouterment and electronics give you a large number of options for good video filming. These devices have automatic functions that perform well in most circumstances, thus providing users with undemanding operation yet solid outcomes

What Are the Benefits to Using a Camcorder?

- The lenses and motor drives give camcorders great zoom capabilities. With such strong zooming options, you can record activities from afar, such as filming business lectures from the back of the lecture hall. What's more, the motor drives make for smooth zooming.

- Camcorders have smaller image sensors than most DSLRs and mirrorless cameras. Even though many camcorders have rolling CMOS image sensors (as are found on DSLRs), the smaller sizes limit the amount of annoying side effects associated with a rolling shutters. As such, camcorders give you more options for filming moving objects or when you have no options but to be on the move.
- Camcorders automate many things for you. They auto-white balance (the process of making sure that white looks the same under different sources of light), focus, adjust for light, and do a good job adjusting for sounds manifesting at different volumes. Press record under automatic settings, and there is a good chance that your camera will do an okay job capturing a scene. The key to using automatic settings is to watch for good lighting on your subject.
- Camcorders come with viewing screens that swivel. You can fold it out such that whomever is in front of the camera can see what the camera is capturing. For one-person production teams or people looking to produce vodcasts, (video-based podcasts), this option is very useful.

Camcorders that produce great footage come with hefty price tags. Price tags can skyrocket to tens of thousands of dollars. These come with a large number of options to optimize your shooting. Indeed, these are powerful tools while filming, however can overwhelm someone not familiar with the settings. It behooves a user to work with the basic options at first and dedicate time to learning how to use the bells and whistles. Knowing that the options exist is one thing. Being able to use them on-the-fly is another.

One of the greatest detractions for reasonably priced camcorders (say, below $1500) is the build of the lenses. Glass quality is typically a little less than that of prosumer (lowest level professional camera) DSLR and mirrorless lenses. What's more, the lenses are typically smaller and assembled with long lens barrels. The long barrels do give the lenses incredible zooming capabilities; however, just like looking through a toilet paper roll, the long thin tube of the lens barrel decreases the amount of light reaching the image sensor.

Another downfall of a mid-level camcorder is their smaller image sensors. Smaller image sensors generally do not reproduce colors as accurately and perform poorly in low-light situations. Low-light shooting may result in grainy shots and poor color reproduction (Image 5.10). Dark shadows appear more speckled with strong purple or muted tones. For films that you wish to have a professional feel, you will likely have to spend significant time color correcting once back at your computer. And in some cases, you might never get the footage to look the way that you like. Vigilance with optimizing lighting while filming is crucial. Shooting in areas with lots of flat light or importing light are steps to improve some of the disadvantages to camcorders.

If you are new to video and are not interested in jumping into complicated and advanced devices, a camcorder is for you. What's more, you can purchase a camcorder for about $100. No kidding! Personally, I think that camcorders start paying solid dividends at around $800.

Small Image Sensor   Large Image Sensor

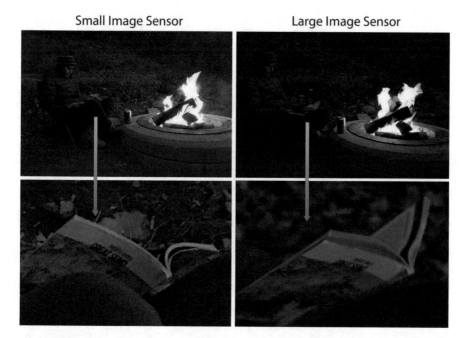

Image 5.10 Here we conducted a comparison between similar shots taken with cameras outfitted with a small and large image sensor. The images shown in the left column were captured with a small sensor and on the right, a large sensor. The lower rows are zoom-ins on small areas within each shot. One can see granularity and color reproduction differences in the coloring of the books pages

Concluding Thoughts on Purchasing Cameras

Determining which camera(s) to purchase is an exercise in harmonizing skill level, filming needs, and budget. We all must take care to work within a balance of these three boundaries as we go shopping. When I step into a store, ready to purchase, it is hard to thoroughly and accurately evaluate the performance of a camera. How can I know the quality of resulting footage when I am only looking through a view-finder? Also, I can feel impetus to purchase quickly so that I can jump into filming. Neither work in my favor. As a result, I like to come into purchases as an educated consumer. Reading reviews on whether a camera is good or not is not good enough. More importantly, I try to understand if items are good at the services that I need fulfilled. If a salesperson makes a compelling pitch to buy something that I had not considered, I take 10 min to jump on my smartphone to read reviews online. Unless you know the background of the salesperson, who knows if what they are telling you about a camera is trustworthy.

Chapter 6
Purchasing Production Equipment: Other Devices

Capturing Sound

Sound quality makes or breaks a film. No matter the budget for your film, most producers and editors place huge emphasis on quality and mood-appropriate sound in their films. I like to think about sound recording as a potent investment in the impact of your film – meaning large improvements can be made to your film for a limited effort in sound recording and editing.

Think about it this way: let's say you're buying a bottle of wine. If you spend $1 for a wine, your expectation is that you will get a terrible wine. Now, let's say that you buy a $5 bottle. You might end up with a wine that isn't great, but leagues better than $1 wine. A $10 bottle might fetch you mediocrity and a $20 bottle, good. However, a $90 bottle might be great and a $100 bottle only little better. Gains are rapid when you go from next to no expense to $10; however, the gains diminish as your budget expands (Image 6.1).

Some people call the relationship of gains slowing with additional expenditure the *law of diminishing returns*. Lots of people will note the difference between a $1 and $10 wine. Few will discern between a $90 and $100.

These principles apply to how you record sound. In other words, if you put anything more than marginal effort to improve the quality of your sound, your film will be significantly improved. These gains heighten impact and information transfer. So never forget to be attentive to sound. If you start out using the microphone built into your smartphone or GoPro, you are drinking the "$5 bottle of wine." A little thoughtful use of this equipment or the addition of an external microphone will quickly take your work to the level of a "$20 bottle of wine." Good stuff!

R. Vachon, *Science Videos*, https://doi.org/10.1007/978-3-319-69512-9_6

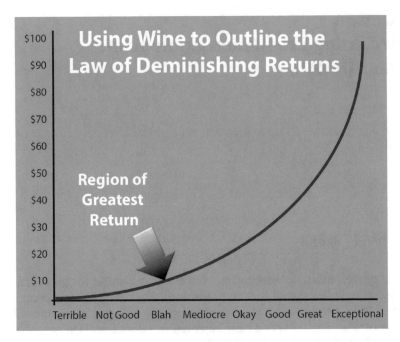

Image 6.1 The *law of diminishing returns*. Wine, as with sound editing, benefits from investment. Paybacks are greatest when efforts jump from no investment to a little. The paybacks gradually become less as you place more and more money and time into it

Types of Sound-Recording Devices

As with cameras, there are numerous types of microphones from which you can choose. Here we will talk about built-in, lapel, shotgun, and portable sound recorders. Each have applications for which they excel, and also have weaknesses. Let's dig in.

Built-In Microphone

What if the only microphone that you have is built into your camera? Typically, film producers think of onboard microphones as fairly dysfunctional for most film production pursuits. This is not a universal rule because some camcorders come equipped with microphones that capture great sound quality (such as shotgun microphones [more on these soon]). In general, though, cameras come with small cardioid microphones. These pick up sound from all directions around your camera. If you need the feel of broader environments, such as wind in the trees, street traffic, or laboratory noise, these are workable. Alternatively, the positioning of these microphones can muddle audio tracks with distracting background sounds and hiss.

Being aware of the problems with built-in microphones is the first step to finding solutions. Your best solution is to buy auxiliary microphones (like those described below); however, you can also use your brain to work with what you have to optimize sound.

Where might you run into bad sound? Built-in microphones are really bad at recording interviews. The key to a good interview is to pick up sound very close to a speaker's mouth, thereby reducing distracting background sound. Built-in microphones are also poor at capturing ambient sound while it is windy. Creating a wind block, with your hand or body, protects the microphone from these disruptions. Since built-in microphones pick up any sound emanating from near the camera avoid rustling your hand over the camera body or don't where clothes that crinkle.

Lapel Microphone

Lapel microphones are very small external microphones, thus unobtrusive, designed to capture spoken word while the camera is positioned at a distance. Most lapel microphones, appropriately, clip onto the front of a person's shirt. This way, the recorded sound is dominated by the individual's voice. Background sounds still exist, however are comparably small. Careful positioning of a lapel microphone, out of a person's breath, delivers the best results.

The cheapest lapel microphone is simply a tiny cardioid microphone on the end of a long wire. Videographers hide these wires within the layers of the speaker's clothing, while the small microphone, on the subjects lapel, is the only visible component. The other end of the wire connects to the auxiliary audio input on your camera. If you spend some time searching online, you can buy these for $20, but $40 or more is common. For a little over $100, you can start to tackle the problem of eliminating the wire leading from the interviewee to your camera by purchasing wireless lapel microphones. These slightly more expensive microphones transmit a signal from a box off of the interviewee's belt (or in their pants pocket) to a receiver attached to your camera. More expensive lapel microphones give you a number of choices of channels upon which to transmit your signal and more power for longer transmission. Also, they are more sturdy so more dependable and last longer (Image 6.2).

Inexpensive lapel microphones transmit on one or a few frequencies. Electric current like that running in your walls or through power lines can interrupt transmission signals. Multiple channels give you options to avoid such disruptions. Plus, multiple channels mean that you can use several lapel microphones to record the voices of several people (one device per person) simultaneously. Each microphone will be set to a different frequency. Note, using more than one lapel microphone during a recording will likely necessitate a *mixer* (a device for adjusting volume levels and combining several sound tracks into one track). Complex, but with certain applications, worth it!

The bottom line? If you are careful with lapel microphones, your sound quality for interviews will be incredible!

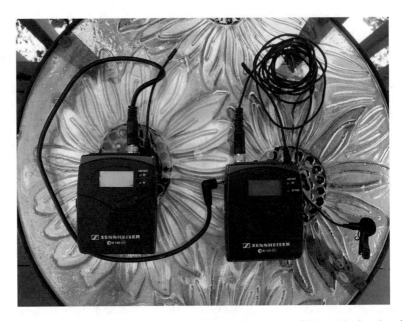

Image 6.2 Lapel microphones are great at capturing a subject's voice. Users snake the wire of the microphone under a subject's jacket such that the small microphone appears through a buttoned shirt (or other kind of garment) near the subject's sternum. Situated so closely to the speaker's mouth, the sound recording is dominated by their voice and not by background sounds. The sound signal is then brought to the camera or other recording device along an extended wire or transferred via small radio transmitters

Shotgun Microphone

A shotgun microphone is a useful microphone for many situations. Uniquely, it picks up sound along the axis of the microphone's barrel, thereby minimizing sounds that might come from the sides. The channeling effect of these microphones means that you can capture good sound as long as you point the microphone in the right direction (Image 6.3).

Shotgun microphones easily mount onto the end of a boom pole or the hot shoe (the clip where a flash connects) at the top of your camera. In the case of recording voices of actors, shotgun microphones are an improvement over lapel microphones for a couple of outstanding reasons. First, audiences won't see a shotgun microphone when it is when it is either mounted on the hot shoe or on a boom, just out of the framing of a shot. Unless well-disguised, lapel microphones are visible, which ruins the feel of authenticity in a scene for some people. Second, lapel microphones are susceptible to disruptions from a subject's breath or rustling of clothes. Shotguns are positioned far enough away from such sounds that these disruptions fall away. What's more, when shotgun microphones are mounted to the hot shoe of your camera, you can film many situations rapidly and still capture great sound. Let's say that you are at a concert and want to get audience feedback

Image 6.3 Shotgun microphones dominantly capture sound along the axis of the microphone barrel, subduing the impacts of distracting noises that arise from the sides

during intermission. A shotgun microphone, directed along the axis of the lens, will easily block out many of the voices from the sides and highlight the spoken testimonial.

The starting price of a shotgun microphone is about $100. I have found the performance of this buying bracket a bit lacking, but with a jump to a few hundred dollars (or a carefully used $150), you are buying a great tool for your quiver. For those hoping to produce a film on a small budget, this cost might burden the purse strings, so perhaps forgo the $1500 dollar camera for the $900 version facilitating investment in a shotgun microphone.

As with any microphone, wind can disrupt a good recording. Other than filming in areas buffeted by the wind, take every action to reduce the amount of wind that blows over the sound-recording end of a shotgun microphone. Most come equipped with a foam windshield that will reduce much of the wind that could make for poor sound recording. Supplementally, you can purchase far more robust microphone wind guards. I have found these additional guards irreplaceable when filming on the run, in hard-to-reach environments.

Part of the nature of shotgun microphones is their clumsy size and shape. They are light, yet stick off your camera like an outrigger. If you truly are running and gunning (shooting scenes rapidly), take care to not torque a shotgun microphone off your camera body. I don't think that most people have too many problems on this front, but they take some getting used to. What's more, always bring an extra set of the rubber bands that are used to dampen vibrations within a shotgun microphone.

Image 6.4 Small portable
sound recorders are
fantastic tools for your
quiver. Focus your search
on those with a high-end
built-in microphone and,
when possible, multiple
auxiliary audio inputs

Portable Sound Recorders

Portable sound recorders do just what their name says. They are compact and move-
able boxes dedicated to sound recording (Image 6.4). Whether you use the microphone on board the portable sound recorder, which is typically of good quality, or
you patch in one or more axillary microphones, these devices are indespensible
backup microphones. What's more they function as small mixers for multiple tracks.

Cool Trick
We quickly mentioned mixers above. Mixers, or small soundboards, are useful for taking sounds from multiple microphones and funneling them down to
one sound track. The sounds coming into each channel can be from very different sources. A great example is a concert where independent microphones
are recording the singer, bassist, guitarist, and drummer (Image 6.5). However,
if you want all instruments to play through house speakers, all need to be combined to one track. If you are running and gunning, with limited time and help,
some portable sound recorders can second as a mixer for a couple tracks.

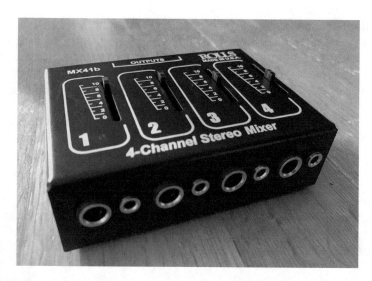

Image 6.5 A portable sound mixer takes sound inputs from numerous sources and channel them down to one track

Camera Stabilization: Tripods

Image stabilization is the bread and butter for producing many quality films. Filming does not necessitate a tripod; however, the stylistic feel of a floating camera (on a shoulder, for instance) can prove quite distracting, so it behooves us to use tripods in most applications. Many people new to filmmaking may not be familiar with the breadth of uses for tripods. They seem so simple, yet read on. They are surprisingly versatile.

Basic tripods start in the $30–$50 range. In very simple settings and uses, these tripods get the job of camera stabilization done. However, they are typically not built with fine-tune adjustments or longevity in mind. For $30 you can step into a product that will hold most prosumer cameras in several positions and, when wielded with an informed mind, will deliver acceptable shots. For example, inexpensive tripods shake and quake in wind, sending vibrations into your camera. These vibrations can ruin a shot. Fortunately, many tripods have hooks located between the legs. Now you can hang weights off the bottom, thereby increasing the heartiness of the tripod and reducing vibrations.

Many inexpensive tripods are also small, which reduces their utility in some filming situations. I would suggest investing in one that rises to 5 feet or greater off the ground. This way you can look through your camera's viewfinder in a standing position. Even more – tall tripods are great for conducting interviews. Be warned though: a tripod at full extension is not likely very stable. They sway and wobble under the heft of a medium-weight camera. Additionally, they offer inexact adjustments to leveling your camera. Extra effort can be made, thus producing useable media. To assure the best-quality footage, give the camera a few seconds to stop swaying and shaking after you press record. Leaving your hand on the camera to reduce the shakes, can be a double-edged sword. Swaying stops, but our hands often produce similarly distracting tremors.

If this $30–$50 tripod is what you are considering buying, here are the device's pros and cons:

Pros:

- They are inexpensive.
- They are extremely light and portable.
- They second as a decent tool for steadying a camera while on the move. More on that coming up.
- They pack efficiently – fantastic for carrying in baggage.

Cons:

- Plastic parts break and are next to impossible to fix when on the move.
- They are not easy to level on uneven ground.
- Their heads or areas where you mount the camera to the tripod, are very hard to microadjust.
- They struggle at stabilizing heavy cameras.
- They have limited options for leg positions and stabilization.

I find that there is a considerable jump in price from the tripods that can be purchased at a Target or Walmart up to the next tier. Two hundred to five hundred dollars fetches much more robust tripods. The legs are strong and may extend another foot longer. While this makes for a more cumbersome carry, once in place, these tripods resist vibration under heavy winds, carry a heavier camera, and facilitate undisrupted adjustments to the camera during filming (e.g., zooming) (Image 6.6).

In order to rotate a camera to capture a wider scene (known as "panning"), it pays to have a tripod head with smooth resistance while pivoting. The head of a tripod is the mounting unit between your camera and the legs. Inexpensive tripods are poor at pivoting smoothly; however, just because you spend more money on a tripod does not necessarily mean that you will overcome this problem. Why? There are a number of tripods made purposefully for still shot photography. Heads dedicated to still photography are often built with a ball joint, allowing for large ranges of position-adjustment. Since photography does not mandate smooth pivoting, ball joints are not designed with such needs in mind (Image 6.7). When you go shopping for a higher-quality tripod that will move gracefully, consider purchasing one with a "fluid head." The "fluid head" does not likely have fluid in it at all – rather, the head will turn on an axis fluidly. These devices are designed for video applications, in that they are easy to level up, tilt, and swivel smoothly.

Image 6.6 Large, more robust tripods afford you stability in your shots, however are less portable than mid-ranging tripods

Here are the pros and cons of more expensive fluid head tripods:

Pros:

- Very stable
- Have longer and precisely adjustable legs
- They are easier to level and microadjust.
- They give more options for smooth panning shots.
- They are made of metal and typically easier to fix.

Cons:

- Their weight make running and gunning more difficult.
- Their prices can strain limitted budgets.
- They are fatiguing and cumbersome to move.

Image 6.7 First, we depict a high-end tripod head dedicated to photography. These provide a large degree of articulation because the assembly is mounted on a ball socket. While these heads are fantastic for photography, they fall short on fluid movement. Second, we show the optimal tripod head for many professional videographers – a fluid head

Devices That Add Movement and Dimensionality

Have you ever been watching television and a scene glides through a landscape? Maybe the camera floated over the Serengeti Plains – almost supernaturally? Or maybe the shot lowered from an upstairs window down to a conversation on a patio on the next floor down. Such gliding shots are useful for establishing a wonderful tone and sense of space. These very dynamic shots, that we will call "active camera", add a high level of professionalism to a movie.

Big-time movies use active camera positioning quite often. They employ incredibly hi-tech cranes, image stabilizing steadicams, and dollies rolling on tracks. Sure, these add amazing perspectives to a film; however, I am guessing that such measures are outside of your budget. Indeed, if these are on your wish list, conduct a search for local companies that might rent them to you. However, let's now talk about some ideas for how you can use or build very functional devices for little cost. They might not have all of the bells and whistles of factory-produced mechanisms, but with careful use and creativity, they can achieve high-quality shots while executing complex camera movements.

Steadicams

Let's say that you want to record a conversation while your subjects are walking. You can track their progression with a zoom lens and faraway tripod. However, what if you want to add more personality and closeness to the conversation, filming beside the moving speakers might do the trick. Unfortunately, filming while walking beside the subjects is extremely hard to capture without generating awkward shakes through the camera. Even if you are walking slowly along a sidewalk, your gate bounces and sways. There are some useful solutions for this problem.

The film industry has developed stabilizing devices, and the best known is called the *Steadicam*. This is a mechanism that straps to a videographers body, yet reduces the impact of a moving body on camera orientation. It facilitates movement over uneven surfaces without significant compromise to the footage being recorded. Although Steadicam is the formal name for a single, very well-designed and specific device, the term "steadicam" has become synonymous with more generic apparatuses that serve the same purpose. If we use the word *steadicam*, with a lowercase "s", we are referring to any device that functions as the Steadicam.

The Steadicam was invented by Garret Brown back in the 1970s and has been heralded as a massive breakthrough for filming. Steadicams and knock-offs are easily purchased online. You might be able to buy one for as little as $500, but don't be surprised when you see price tags in excess of $1000. Expensive, yes, but the outcomes bring immense value to moving shots. So, are there more frugal options out there?

One economical alternative uses the principle of hanging weight beneath a camera to even out unstable shots. They typically mount beneath a camera, providing the user with a handle and counterbalancing weights serving as ballast. These can be purchased for as low as around $100, but with a little time and a basic workshop, you can build your own for $30 (Image 6.8).

There are a number of ways to build a basic steadicam, and what I describe below is one very practical method. The process starts with a visit to your local hardware and sports store and ends with a couple of hours of assembly time (half an hour if you are proficient). Since the building is simple, all that you need is a lab, garage, or kitchen table; the most basic tools; and an understanding of how to use them.

Image 6.8 Inexpensive steadicams are a good first step towards stabilizing your camera when filming on the fly, or running and gunning

Here is the list of parts that you should purchase from a hardware store:

- 2, 10″ segments of ½" diameter black iron pipe (A)
- 1, 6″ segment of ½" diameter black iron pipe (B)
- 1, T-bracket for ½" diameter black iron pipe (C)
- 3, end caps for ½" diameter black iron pipe (D)
- 2, 1 ½" long ¼" diameter machine bolts (E)
- 3, lock washers for ¼" bolts (F)
- 4, washers for ¼" bolts (with diameters greater than 1″) (G)
- 1, ¼" diameter wing nut (H)
- 2, ¼" machine nuts (I)

Stop by a local sports store for a counterbalance weight. The amount of weight that will serve as ballast will depend upon the weight of your camera. The more your camera weighs, the more weight that you need for ballast. Unfortunately, the added weight means that heavy cameras become even heavier. In general, I would say that a 2 ½ pound barbell weight will serve to balance a very simple DSLR or small to mid-sized camcorder. A 5 pound weight is more appropriate for larger cameras. The inside diameter of the barbell weights should be 1″. FYI, if you think that you might want to switch between a light and heavy camera setup, you can purchase two, 2 ½ pound weights. That way you can remove a weight for a lighter setup. If this is an option that you like, note, this means that you will need to buy only one, 1 ½" long, ¼" diameter machine bolt and another 2 ½" long, ¼" diameter machine bolt, from the hardware store.

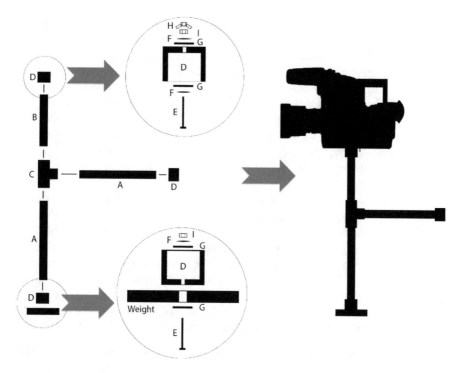

Image 6.9 Assembly of a do-it-yourself steadicam

The assembly of the steadicam requires a vice grip wrench (or table-top vice), drill with a ¼" bit, and pliers (or monkey wrench) (Image 6.9):

- The first step is to modify two end caps to become (1) a camera mount and (2) a mechanism to cradle the ballast weight. Drill a ¼" hole through the center of both end caps. This can be most easily accomplished by locking each cap into a vice, so that the cup is pointing upwards. If the bottom of the cup has curvature, place the drill into the center of the cup and drill downwards. The curvature will force the drill point into the center of the cap. Otherwise, do your best to keep the drill bit centered (a center punch may help to start this process). A consequence of an off-center hole is difficulty placing a washer down into the well of the cap, later on.

- For the assembly of the camera mount, thread a lock nut and then a flat washer over one of the bolts. Then pass the bolt through the cap, so that the head of the screw is nestled into the cup side of the cap and the threads are coming out the other side. Now, feed another washer and lock washer over the threaded end protruding out of the cap. Screw down all of these pieces tightly together with one of the nuts. Lastly, thread the wing nut over the remainder of the screw such that the wings are facing the cap and the flat end is directed towards the threaded end of the bolt. There should be several threads remaining on the bolt. When the whole steadicam is built, you will screw the base of the body of your camera over these remaining threads. The wing nut will then serve to immobilize the bottom of your camera on the device. FYI, it can be challenging to keep heavier cameras from swiveling on this style of do-it-yourself steadicam mount.

- To construct the weighted end of the steadicam, begin with the drilled hole through the other end cap. Opposite to the camera mount, a bolt will be fed through a barbell weight first. A washer on the end of the screw will keep the bolt from passing all of the way through the barbell hole. Now pass the threaded end of the bolt through the top of the end cap. With the threads now pushing out the cup of the end cap, tighten down a flat and lock washer with a bolt. The barbell should now be fixed tightly to the end of the cap (for two weights you will need a 2 1/2" bolt).
- Now is the time to connect all of the major parts. Tighten down the last (third) remaining end cap onto the one end of a 10″ long black iron pipe. The other end of this pipe will fit into the hole of the T-bracket that does not have a mirrored other end (meaning screw it into the base of the "T," not one of the top ends). Fix the 6″ and second 10″ iron pipe to one of the unoccupied ends of the T-bracket. The 6″ pipe will be at the top end of the steadicam, while the 10″ pipe, without the cap, is directed downward. As such, tighten the cap that is fashioned to be the camera mount to the unoccupied end of the 6″ pipe and the weighted cap to the unoccupied end of the 10″ pipe. You are done!

How do you use a steadicam assembly? Thread the base of your camera onto the steadicam's camera mount screw. Tighten the wing nut snugly against the base of the camera (with your fingers). This will keep the camera from rotating. As you use the camera, be wary: the wing nut can come loose. The horizontally oriented iron pipe should be extending either perpendicularly to the left or right of the direction that the camera is pointed. The position of the steadicam is such that one hand holds most of the weight onto the 6″ pipe while your other holds, stabilizes, and steers

Cool Trick #1
The steadicam is typically used with the camera mounted to the top. Unfortunately, the positioning of the weight and the orientation of the camera in relation to your arms makes it very difficult to get low-to-the-ground shots. When would low-to-the-ground shots be useful? Perhaps, following behind walking feet or zooming over the surface of wild flowers. A simple solution to acquiring stable, low shots is to upend the entire steadicam assembly, such that the camera is upside down. You can even take off the weights! The camera now becomes its own ballast. Indeed, any shots that you record will now be upside down. Fortunately, this can be corrected in the editing studio. Editing programs give you the option to flip your shots vertically, thus quickly solving this problem.

with the horizontal 10″ pipe. Fortunately, this design lets you reduce fatigue over long days by turning the steadicam 180° – that way each of your arms can steer and hold the weight. All that you need to do is change the orientation of the camera by 180° as well.

Most users suggest that you are the most critical element in making a steadicam produce good results. Bend your arms such that they act as suspension for the assemblage. Practice gliding while you walk. Keep your shoulders upright, because slouching over long periods of time can lead to back pain.

Cool Trick #2

Another modification helps with challenges associated heavy cameras fixed to do-it-yourself steadicam. In some cases, heavy cameras do not secure well to one ¼″ bolt and spin on the steadicam head. Consider attaching a larger platform for your camera, thus adding more friction against rotation. The old head (with the wingnut) can be replaced by a "floor flange" that threads onto the top end of the 6″ pipe. The flange provides a large platform from which a plank of wood (3/4″ plywood) can be attached (likely with four 3/4 "wood screws"). Before the plank is screwed in place, make certain to secure a ¼″ threaded bolt through it, such that ½″ of thread rises above the wood surfaces. The end of the bolt will later be threaded into the bottom of your camera body.

Steadicam Substitutes

Let's say that you don't have your trusted steadicam on hand. What can you do with the assets at your fingertips? Lots! For example, collapse the legs of an inexpensive tripod and suspend it beneath the camera. Cradle the camera or tripod head gently in your hands such that the additional weight of the tripod serves as ballast.

What if that does not give you enough control of your camera? Try tucking two legs of a collapsed tripod into the waist of your pants. Your legs and core add immense stability to your shot. Wherever you might need a shot, span your legs into a stable position, with shoes separated by a couple feet. Take the weight of the camera through your belt and support the camera with your hands. The camera will likely be very close to your face. To reduce breathing sounds, and a rise and fall in the framing (associated with breathing) consider gentle breaths or even holding your breath as you film. With practice, you can walk with this construct (Image 6.10).

This method works best when you use a wide-angle lens. That way, small trembles through the camera do not translate into big disruptions in footage. Zooming in can result in shaky and, often, irreparable clips.

Image 6.10 One can film
fairly stable shots when a
medium-sized tripod is
collapsed and two of the
legs are tucked into your
beltline. This affords you
movement over difficult
terrain and smooth pivoting
about your hips

The anchoring of the tripod through your core is also useful for executing somewhat smooth pan shots. Rotating your torso over your legs provides a smooth pivot
point as you brace the camera in the same position that was earlier described. I
would emphasize rotating much more slowly than you would ever imagine.

Side Note: I share ways to simplify stabilization of shots. In general, these options
work best for those just learning how to film, or are working on a shoestring budget.
Let's step in the direction of more expensive options for a moment. What if you are
on a boat and you want a film clip to keep as level to the horizon as possible?
Gimbals are built for these applications. Gimbals typically position a camera within
a series of concentric rings rotating at 90° to each other. Complex gimbals have
three rings correcting for the pitch, roll, and yaw. Some gimbals do not center the
camera within rings, but they serve the same purpose.

Sliders

The first time that I saw or recognized the use of a slider was in a scene that slowly rolled from behind seated fans in bleachers into the aisle that descended to a baseball diamond. I felt like I was sitting in a lively baseball game. Gripping!

A slider is a glide-track upon which you connect your camera body. Your camera is fixed to the tracks with a small rolling cart. You then gradually push, by hand or with a motor control, your camera through a scene. In order to articulate your camera to unique angles, it is common to attach the camera to the cart with an adjustable tripod head.

While a slider can be positioned on the ground, the applications for its use grow when it is fixed to a tripod with a strong head (thus allowing you to change its angle and orientation on the fly). A well-positioned tripod and slider can effortlessly move your camera over strange terrain or through a maze of analyzers. The tracks can be as short as a foot and a half and up to tens of feet (no longer fixed to one tripod). Personally, I have a four-foot slider tucked into an old snowboard bag that is ready to go on a moment's notice. This slider is easy to move, fits into the trunk of any car, and provides ample track upon which to slide your camera. However, an immense amount of material can be shot with half that distance. Indeed, my two-foot slider is easier to wield when I am depending upon one tripod to lock the angle of the track (Image 6.11).

Prices for hand-driven sliders start at $30. I would warn against using sliders as inexpensive as this, unless you keep it well tuned. Why? The primary function of a slider is to provide a smooth glide. Wheels on inexpensive sliders become obstructed and make for uneven movement. In my experience, $300 is the gateway price to sliders that consistently deliver great moving footage. The price of sliders jumps once again when they come equipped with a motor drive that provide a very slow and smooth glide. Unfortunately, these benefits increase the cost and complexity of transportation and assembly.

Sliders work best when you can move your camera evenly along the tracks at whatever speed that you want. Take care to practice even glides. Each person has their own solution; however, I often drag a couple of fingers from the same hand holding the camera/cart along the rails, creating even resistance. Alternatively, steady your arms against your side and use the bulk of your body to move the camera with inertia.

The simplest shots are horizontal. Whether your slider is leveled on the ground, a table, or a tripod, simply move your camera sideways across a scene. The photography tripod head (ball heads are the best) mounted to the cart gives you many options for locking in the direction of your lens. Warning: Whatever your method for steadying your slider, take extra time to make certain that this placement is secure. Moving cameras can destabilize precariously positioned sliders.

Image 6.11 A two-foot long video slider. A camera is usually attached to the slider cart with a ball-head mount, and the slider is attached to a stable tripod with a fluid head. The integration of robust tripod heads allows you greater degrees of movement and creativity with your shots

Here are a few imaginative techniques for making stunning effects:

1. Orient the tracks of the slider to line up with the direction of the lens. Use the cart to close or open the distance between the camera and elements in the scene. This will allow you to navigate the camera through objects like dishes on a table or plants in a greenhouse. Be sure that viewers can't see the end of the track when your camera is farthest from the elements in your scene. Sometimes, it is provocative to have elements of varying importance move in or out of focus as you move the cart. For example, start with an element out of focus, yet as the camera rolls towards it, the object becomes resolved. The gradual revealing of an element creates great tension in a storyline – realization or hard-won clarity. Oppositely, pulling a shot out of focus by the end of a clip leads to a feeling of muddling or loss (Image 6.12).

2. Sliders have applications outside of the horizontal domain. Try tilting your slider, such that you, say, move from under a table to people huddled over a computer. If you position your slider such that the camera (and thus your viewer) emerges from a closed space, the audience may feel more engaged with the surroundings.

3. Loosen the ball joint of your photography tripod head (mounted to the slider cart and camera) so that you can pivot the head freely. As such, you can articulate your camera to be directed at a closeup object as the cart moves across a scene. This is accomplished by keeping two hands on the camera as the camera rolls along. Your hands will twist the camera so that the lens is always pointed at the object of interest. I find that this effect is most potent when the objects of interest

Image 6.12 One can use a slider with the camera oriented in near alignment with the path of the slider rails. This makes for the unique effect of moving through subject matter. One must be careful not to capture the rails in the shots

Image 6.13 A more advanced application of a slider includes loosening the ball head attached to the bottom of your camera. This way, as you slide the video cart over the rails, you can keep the framing of the camera fixed on a passing object

are about 1–5 feet away. Any further than that and the impact of the effect becomes lost (Image 6.13).

4. You can also create shots by tilting the tripod head, holding the tripod to the slider. That way, you can change the angle of the slider while you pass the cart along its path. Maybe the scene is tilted at first, gradually levels out, and then tilts

the opposite direction. It can simulate a feeling of being on a boat rolling in the waves. A fluid head on top of your tripod works best for these purposes.
5. Get crazy and try using as many of these techniques as you can in one scene. Spin, slide, and tilt! Why? Because when these shots work out, they are very exciting! Practice with any objects that allow you to hone your skills before you are put to the test on a project that really matters. Warning: These shots are wild; thus, there is a lower chance of success. Make sure that you have the time to give these methods a healthy try. What's more, it is helpful to first capture simpler footage that will also fulfill your needs. Capture necessary footage before getting fancy.

A Two-Ton Slider: Car Glides

If road conditions are just right, you can upscale your slider to the size of a car. You can float your camera through trees or in front of draft horses preparing for a sleigh ride by creeping your car through different spaces. Let's call this a *car glide*. Car glides benefit from a few very simple measures that deliver useful and interactive shots (Image 6.14):

- The key is to secure your camera on your window frame, doorframe, dashboard, or on the roof; however, adding some suspension, like fingers or even a folded towel, helps reduce vibration.
- In general, I use the car glide to shoot broad panoramic/landscape shots, although well-placed, smooth driving surfaces open opportunities to situate quite close to the elements that you hope to film. Particularly when looking outward from a side window, you can sweep closely by post office boxes, signs, trees, and people (if willing).
- You would not believe how little road that you need to execute a car glide. I have filmed car glides with a length of road as short as 5 feet. I find that short car glides benefit from at least one foreground element to give perspective on changing spatial relationships. If you are shooting landscapes, without foreground elements, the framing of faraway elements will change little with a short glide, thus ineffectual. Foreground elements juxtaposed against faraway scenes produce a greater feel for the dimensions of a space.
- It is much easier to film these scenes with a car that has an automatic transmission. With an automatic transmission, you can smoothly start moving from stop. A light flaring of the brakes easily maintains slow and constant speeds. Cars with manual transmissions are harder for controlling slower speeds. In my manual car, I have turned to slightly sloped hills for a gradual roll.
- Open your lens to a wide-angle (as opposed to being zoomed in) setting. It is practically impossible to find a road that is as smooth as a railed slider; thus, any

Image 6.14 Car glides are a powerful way to capture footage passing through scenery. It is important that you find smooth roads. If necessary, pass a stability filter over your footage once back in the editing studio

car vibrations are communicated into your shot. A wide-angle setting reduces the impacts of the small disturbances that will inevitably disturb your car glide. To think that you can zoom in is wishful and typically unrealistic.

Cool Trick

Large-scale production companies shoot cameras from on board rapidly moving trucks. These are great for filming car chases. The technologies that stabilize these shots are extremely expensive. For most of us mortals, options to use this gear are beyond our budgets. That said, what if you think that filming from a moving car is critical to your story, yet no matter how you search for a smooth part of road, or attempt to suspend your camera, vibrations and bumps render the footage unusable? Modern editing programs have applications that remove an enormous amount vibration that plague otherwise very useful clips. For example, the video editing program Adobe Premiere Pro uses the effect called "Warp Stabilizer." The framing and resolution of shots are typically slightly reduced in the process, but can take sadly unworkable footage and bring it to life. Even if you do not have a program that has such options, place the clips that need post-production (editing) help with vibration into a folder that you bring to friends who can help you. A savvy editor can import the video files, place a hasty image stabilization filter over your clips, and export them as new files fairly quickly. Your problems might be solved!

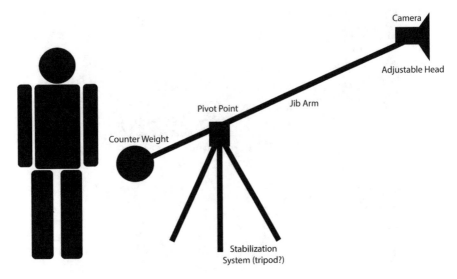

Image 6.15 Generalized jib assembly

Jibs and Cranes

Hydraulic cranes and booms transport cameras and wiring over incredibly difficult, if not impossible, terrain. In videography, a long boom arm with the camera fixed to one end is called a *jib*. Jibs can swoop from very low angles to a bird's eye view or vice versa. As jibs swoop over large spaces, videographers can view what they are filming from the safety of a monitor affixed to the controls below. Modern phone apps are fantastic for viewing what your camera is recording in real time. Jibs are great tools for giving a professional feel to a story and can wow your audience into wanting to see more of your work.

In general, a do-it-yourself jib includes a long arm (5–12 feet) that extends away from a stabilizing point on the floor (often a heavy-duty tripod). The arm is designed to swing up and down, and pivot in a circular manner around the stabilizing point. The far end of the jib arm is where you attach your camera. As such, when you move the jib arm, the camera has an enormous range of area that it can cover. Sometimes the camera is locked in place at the end of a jib arm, but many jibs use levers to change where the camera is pointing. Alternatively, some are designed to maintain a camera's orientation, such that no matter how you change the vertical angle of the jib arm, the camera is always pointed in the direction that you initially set. The opposite end of the jib arm extends beyond the stabilizing point to a counterbalance weight. This reduces the effort that you need to lift the camera at the far end of the boom (Image 6.15).

Just like most of our conversations so far, with some searching, you can find places that rent jibs. Unfortunately, even rentals may be beyond your budget. Fortunately, there are do-it-yourself options! Be warned: Unlike building your own steadicam, constructing your own jib takes quite a bit of time and some skill.

There are a lot of jib ideas and instructions online. As you search the web for a great design to build, it is important to know might not find one plan that will suit all of your jibbing needs. As such, different designs will use different materials and emphasize special functionality such as smooth swiveling and control of the direction that the camera is.

What size do I look for? My ideal jib combines versatility for executing the shots that I need and enough simplicity to keep me motivated to bring it to a shoot. I want to show up in the field or in a laboratory, take 15 min to assemble my jib (with practice), and get filming. As such, portability of a jib is an extremely important quality. Rigging arms any longer than 8 feet (tip to tip) can be unwieldy. They do not fit into elevators or cars, and snag on tree branches. I find that I don't use jibs that complicate my day; thus, elegant simplicity reigns supreme …for me.

Unmanned Aircraft Systems: Drones

Unmanned aircraft systems (UASs), commonly called drones, add a fantastic dynamism to films. One can now purchase UASs with a high-definition camera mounted to their belly, such that the cameras record what a pilot would see if they were looking out of a cockpit (or below their feet, more accurately). These helicopter-like vehicles are usually outfitted with four or more rotor blades. The numerous blades make for smooth passage and agility. There are many applications for UASs ranging from following cyclists to showing a vast river from high over the treetops (Image 6.16).

The first thing that you need to know about UASs is there are restrictions to where they can be used. Since laws and controls are changing, it is your responsibility to figure out what licenses you need and whether the desired applications for your UAS are legal. What could be a few examples of limitations? UASs must be visible to the pilot or spotter at all times. Don't fly a UAS above 400 feet and never fly a UAS near other aircraft or airports.

Knowing their predicted flight times is part of planning for filming. How many flight patterns can you run? How far away can you fly? The life of a UAS battery is, in part, defined by temperature, wind, and altitude. Lower temperatures reduce battery life, and higher altitudes mean that a UAS has to work harder for lift and maneuverability. Why? There is less air off which to push. As a result, batteries drain faster.

Introductory UASs fetch as little as $100. These devices can capture some very fun footage; however, they come with substantial limitations. For example, they usually do not film at resolutions greater than 720p. Additionally, the camera is locked in beneath its body. Wherever the drone points, the camera films. What's more, you cannot monitor what the camera is recording, their batteries are short lasting, they have small distance ranges, and they are less steady in flight. As you spend more money on devices dedicated to filming from above, the aircraft themselves can stay in the air longer and travel much farther and faster. Some have GPS onboard. Others have software that reduces the risk of them running into objects. Even more exciting, their cameras can be articulated in several directions during flight. To complement this

Image 6.16 Unmanned aircraft systems, or drones, are multiple-blade helicopter-like vehicles. They are adept at carrying small, yet very capable video cameras over broad landscapes. They are powerful tools requiring flying and operation expertise

quality, you, the pilot, can see what the camera is capturing either through your smartphone or from a separate screen connected to the controls in your hands. Expect to spend over $500 if you are looking for a UAS with such accouterment.

Learning to fly a UAS goes well beyond understanding what the levers and buttons on your remote control do. In order for a UAS to capture what you are hoping for, you have to learn to think like you are its pilot. This might seem like a very simple statement, but it's profoundness will sink in the second that you start to fly with filming in mind. UASs have a front and a tail (usually indicated by colors and lights). As a pilot from the ground, you have to know which way the UAS and thus, camera is facing. Awareness of the UAS's orientation informs you on which way the UAS will fly when you toggle forwards, backwards, or to either side. What if you then want to maneuver left? If the UAS is flying away from you, turning left on the controls moves the device left. Oppositely if it is moving towards you, it takes some time to intuit that you will have to toggle right. Indeed, your relationship to your controls improves when you can see what the onboard camera sees and confirm orientations.

A number of experienced UAS pilots agree: the gains of honing keen flight intuition before you want to use it for a critical shoot are great. Why chance it if you hope to fly in compromising locations (such as over water, in the wind, or among trees)? Some speak of increased anxiety once the device has gone airborne, especially when they were learning how to pilot their device. Movement becomes frantic and their brains stop thinking clearly.

How do you practice thinking like your UAS? Begin your piloting experience with simple exercises and once honed, move on to more complicated lessons. For example, simply practice taking off, leveling off at a given height, and then returning to the ground where you started. Several times. Sound simple? Try it and learn the sensitivities of your device or how nature throws kinks into the works. Winds, no matter how light, will draw your craft to one side or another. How do you compensate for this wind to land your UAS back where it started?

Let's make things more complicated. Lift off and then spin the device to face left, right, and backwards (meaning its front faces you). From these different orientations, learn how to maneuver the device towards objects on the ground, like a far-off tree (don't get too close... simply move in the direction of the tree). Then try these exercises high above the ground. To improve your game, then try to maneuver your UAS around an object from these different orientations. The key to good filming while flying a UAS is smooth movements. Rough transitions can quickly ruin a lofty and often ethereal bird's eye view.

Cool Trick
Higher-end drones are built with sophisticated leveling and anti-vibration gimbals. Their purpose is to make for smooth footage during irregular flight patterns. Fortunately for you, you can use this technology in your hands. Literally. Depending upon the device, you can turn it on, hold it in your hands, press record, and move the UAS to point where you like. The gimbal on board the UAS will smoothen out walking or hand movements in ways that steadicams or jibs cannot. Unfortunately, you might need to have the blades spinning at an idle, which creates unnerving background sound. As such, this trick might not work for recording moving conversations (like people hiking and chatting together).

Lighting

Adjusting Your Camera's Settings

Entry-level videographers often lean exclusively on native light. If they are indoors, they use the lighting hanging from the ceiling. If they are outdoors, they use the sun or streetlights. It keeps workflows simple. Why do we come to rely on sundry sources of light for quality shots? The automatic settings for most modern cameras do quite a good job compensating for the light levels. Additionally, they sense the light spectra (tints) hitting the image sensor and compensate for the quality of light. This is called "white balance." For example, fluorescent lights often bathe rooms in a more blue color, while incandescent light is much more yellow. As such, cameras executing white balance compensate for these differences to make whites truly white. It is unfortunate that some people take a camera's white balancing for granted, because the qualities of white within a shot is a very powerful tool, lending mood-enhancing light to your films. Think about it: without white balance, shots of baby bear cubs under fluorescent light would seem slightly blue, sterile, and emotionless Viewers won't get the warm fuzzies from such shots.

Depending too much on the automatic settings on your camera can give you a false sense of security in believing that the processor in your camera will solve all of your lighting problems for you. Indeed, camera electronics are rapidly improving, but they are not cure-alls. How do your shots benefit from taking more control? Small camcorders come with small image sensors and less sophisticated circuitry. Two conditions prey upon these issues and make for poor shots. (1) Low light: entry-level cameras are notorious for their inability to reproduce colors in low light. Give it a try. Low-light shots appear very granular and black colors take on a purple hue. This is illustrated in Image 5.10. As such, if you have a choice, avoid filming in dark settings and change locations or use supplemental lighting. (2) High contrast: many inexpensive cameras (with small image sensors) struggle with highly contrasting light, which means that details are rapidly lost in shots characterized by areas of intense light and dark. When filming in bright sunlight, an inexpensive camera will do its best to adjust to the amount of available light, but details in the dark or light will likely be lost to under- or overexposure (Image 6.17a).

With a little knowledge and creativity, many of these limitations can be managed.

As you will find, time and time again, your strongest tool is your mind. Here are a few very simple approaches for creating better lighting scenarios:

- Avoid having your interviewee's face bathed in sunlight. Such intense light is overwhelming, reduces dimensionality on a face, and tends to make the subject squint which is distracting (Image 6.17b).
- Alternatively, seek flat light. This can be as simple as standing in the shadow of a building (as long as it is not too dark). It can seem counterintuitive to step out of the light into the shade, but it works wonders to flatten the light on your subject. Flat light decreases the contrast, such that your camera will pick up the details on an interviewee's face, which more effectively conveys expressions (Image 6.17c). What's more, it reduces squinting and allows your automatic light meter settings to do its job.

The last solution may seem simple, but does come with its own complexities. For example, be wary if your subject is in the shadows but the background is still bathed in bright light. Your camera may try to light meter on the bright background, and as a result, your subject may appear very dark. (Image 6.17) Why? Because your camera's automatic settings "think" that your subject is brighter than what you'd like

Image 6.17 (**a**) Underexposure of subject matter can occur if the automatic light adjustment settings of a camera lock in on brightly lit backgrounds. (**b**) Be careful to keep subject matter out of high-intensity light. This can lead to shots with distractingly high contrast. In the case of interviews, interviewees may squint and deep shadows form under their eyes. (**c**) Seek out flat light for best actio results

Image 6.18 Portable light reflector, also called "bounces"

and takes action to darken the shot. Adjust so that the bright light is no longer in the background, or make manual adjustments on your camera. The adjustment that may solve this problem is explained later (in the chapter on filming).

Reflectors, aka Bounces

How can you use light from nearby to illuminate where you hope to film? Perhaps you are filming in the shadows of a building or conducting an interview with only downward-pointing ceiling lights. Consider reflecting the light from one place to another. For as little as $10, you can purchase collapsible, transportable light reflectors (also called 'bounces'). Some are meant to be held by filming assistants or propped up on a chair, and others can be locked in place with stands (Image 6.18).

Importing Light

What kind of dedicated photo and video lights can you purchase? Firstly, if you are filming subject matter that is quite close to your camera, you can purchase LED light arrays that mount to your camera's hot shoe or over your lens (Image 6.19a). Because of their size, their applications are limited. For wider and more expansive shots, it may be more appropriate to consider larger light sources. Studio lights generally center a light source within a dish-like cone. The cone redirects and focuses light. Likewise, they come equipped with a translucent shield in front of the bulbs to soften the light. More modern studio lights may be an array of LED light bulbs, rather than one, producing less harsh light.

Image 6.19 (**a**) Portrait lighting can be mounted to the hot shoe on your camera. These are great for closeup shots. (**b**) For a small investment, one can purchase lighting designed specifically for video and photo usage. They are light and portable, though sometimes not durable (on set for Flipping 50). (**c**) Shop lights, dragged out of your garage or purchased at most hardware stores, can help in a pinch

Tripods or use the adjustable legs that come with most lighting packages are very helpful. Within seconds, you can angle the light in a number of directions and raise it to variable heights. Likewise, these may have refractor panels above, below, and off to the sides of the bulbs. Prices for studio lights begin at under $100 (Image 6.19b). For spaces dedicated to film purposes, these lights are indispensable. If you wish to devote a heap of time and effort to portrait/interview/studio-style videography, investing a little more money into these lights goes a long way.

How about simpler a solution? If you are filming indoors and brightness is inadequate or spaces are cloaked in darkness, consider a $30 or less halogen light from the hardware store (Image 6.19c). The light that they produce is very intense and later in the book we talk about methods to disperse this.

Survival Kits

As you film, your senses must be alert for changing light, background sound, battery life, and permissions and locations for interviews. Unseen challenges can throw a monkey wrench in everything. I have been confronted with broken pieces of gear, screws falling down cracks, and inclement weather. Additionally, filming is about you showing up with your A-game. As such, always come prepared to solve problems. A survival kit is always part of my kit. I typically have a small canvas sack tucked in a far corner of my camera bag or in a separate backpack that also holds my water, extra layers of clothes, or paperwork. Here is what else I carry:

- Multi-tool (which I have to take out when I board planes)
- Small roll of duct tape or gaffer tape
- 8 feet of thin chord
- Pen and tiny pad of paper
- Two twist ties
- Sunscreen
- Two energy bars (passing one to an interviewee or assistant is great for making friends and closing a day of filming with a smile)
- Pen flashlight
- 2 AAA and AA batteries
- Any specific tools that you know might help along the way (say, an allen wrench for tightening scaffolding)

Camera Bags

Choosing a camera bag takes a fair bit of thought. Before figuring out dimensions, configuration, and number of bags that you hope to buy, be aware that there are many inexpensive camera cases that look good and cradle your gear nicely, but barely last a hard day of filming. The zippers, stitching, or even the material are prone to fall apart after a remarkably short period of time. If you have a very limited budget or are only going to film occasionally, purchasing these cases might make sense. If this is your situation, the best deals are found online at the massive camera sales warehouses where you can buy comprehensive camera packages at seemingly great savings. However, I turn to bags that have glowing online reviews from numerous people.

With the conversation of quality covered, now you can think about the characteristics of a bag that will suit your needs. I would start by identifying the intended uses for your camera gear. For example, are you only going to be filming in a studio? In this case, a camera bag built for on-the-go shooting is not what you need. Instead, look for a case that is easy to pack, organize, and store. If space is not an issue, I sometimes store the separate elements of my kit (camera, audio, lighting) in individual cases for quick management. I store them in an easy-to-access closet or on bookshelves whereby grabbing one bag/case will not drag the others off the shelf. Simple and organized.

I store my expensive gear in robust cases. It is worth every penny protecting my carefully purchased quiver with dependable hard cases. Frequently I turn to a company like Pelican for durable and watertight containers. There are less expensive options, but I have rarely found lasting performance with these.

I often take my quiver on the road to sundry locations. These settings vary from offices, to the woods, to high mountains, to laboratories. Gear separated into several cases complicates my day and increases chances for confusion. Typically, the smaller and simpler my setup, the more time that I can spend capturing the footage that I need. The balance that I have struck has come from years of experience in my particular field. What's more, users learn to work with the package that they have chosen and stuck with for some time. Constantly changing quivers creates confusion. To be unfamiliar with my quiver reduces efficiency and may result in subpar footage.

In a moments notice I can refine my quiver down to one primary bag for filming on the go and a smaller secondary bag. My primary bag never leaves my side. Its size reflects careful decisions about what gear I believe that I will need coupled with adequate supplemental gear (such as batteries and memory cards). What's more, the bag has to fit where I plan on going. With the one bag in hand, I seek the safest option to transport my auxiliary bag.

My large bag is a rectangular backpack, with protective foam around the outside and foam inserts so that I can personalize how I compartmentalize the larger internal region. The approximate dimensions are 20 inches tall, 14 inches wide, and 10 inches deep. This bag is comfortable on my back for long periods of time and fits easily into overhead compartments of planes (I fly for projects quite often). The spacious interior affords two DSLR camera bodies, three to four lenses, camera batteries, and a lapel microphone kit. If I drop one lens from this grouping, I can fill in its space a compact shotgun microphone. What's more, the smaller pockets give plenty of options for storing microphone batteries, memory cards, energy bars (to stay lively), and accessories. The big deal is the side zipper pocket that slots in a laptop

Image 6.20 My large camera kit includes two camera bodies, three lenses (or two lenses and a shotgun microphone), a lapel microphone kit, and extra batteries and memory cards

computer. When necessary, side lashings hold a collapsed tripod. Should I choose to shoot with a camcorder, the DSLR lenses and camera bodies can be yanked from the kit, the dividers repositioned and the camcorder put in place. Whether on the streets of a developing country or fighting jet lag on the floor of Chicago O'Hare airport, this backpack keeps my quiver beside me. All the same, once I arrive on sight, I can rapidly ready my gear (Image 6.20).

Another kit that I turn to is quite a bit smaller. It is an over-the-shoulder soft carrying case measuring in at about 12 inches tall, 10 inches wide, and 5 inches deep. The main compartment, accessed from the top flap, has room and dividers for one DSLR camera assembled with my most often used lens (24-105 mm), one more lens (like my telephoto), and one set of lapel microphones. If I work hard with space management, I can squeeze one more set of microphones or a small backup camera into the pack. There is also ample space for batteries, wires, and memory cards. This bag is built for the war. It is rough-and-tumble and slides under the radar of people who might be sensitive to media coverage at an event (Image 6.21).

Where Should You Purchase Your Equipment?

You have researched your ideal quiver of gear. Online searches give you a good feel for how much things cost. You can price-point everything that you need and build a comprehensive list on paper. And then on the day of purchase, most people end up spending more than expected. For example, you may see that your camera comes complete with a battery or memory cards. However, sometimes the stock

Image 6.21 My small camera kit includes one camera body, two lenses, a lapel microphone kit, and extra batteries and memory cards

accouterments are barely enough to get a job done with certainty. In this light, I try to purchase everything that I know that I will need as opposed to what I think that I can manage with some creativity.

There are several choices of where to look for purchasing the elements of your quiver. On the fly, you can head over to your nearest *Target*, *Walmart*, or *BestBuy* and come away with a fantastic introductory camera and warranty. If you time it right, sales can bring prices down to within a spitting distance of those found online. Additionally, you can typically return defective or poorly chosen components with few questions or hassles. Unfortunately, these stores normally come up short with higher-end cameras, lighting, sound, and image stabilizing options. Fortunately, local purchases mean that you could be experimenting with your camera an hour after walking out of a store with your purchases.

Perhaps you are lucky enough to have a mom-and-pop shop nearby. I have found that prices at these shops are a wee bit more than those found during sales at *Walmart*; however, most have significantly larger selections for higher-end purchasing needs. Sometimes a little more expense at the outset of your exploration into cameras and video will save you big bucks and time down the road. Most importantly, befriending shop managers provides you with a priceless resource – their bounty of information and assistance. Whether you are on the fence about stepping up to the next tier of cameras or need advice on how to sync your recently-purchased camera to your phone, the wealth of knowledge and interest crammed in the brains of these folks is indispensable.

How much have you researched your camera and audio needs online? Chances are advertisements popped up for great deals on the exact gear that you need. Are these healthy leads to follow for purchases? Typically, these advertisements are linked to a distribution warehouse where those who cover your call or package your shipping are not knowledgeable about your needs (however, this is not always the case). If you know exactly what you want and need, this approach to online shopping can fetch you great gear for very little money. Additionally, some of these stores give you very inexpensive options for purchasing accompanying accessory bundles with your camera. Tripods, memory cards, lenses, camera bags, and countless other goodies will land on your doorstep in just a few short days.

Be aware though. Online camera distribution companies are sometimes keen on hooking you into buying more on supplemental sales, and thus the prices and quality of these often third-party additions can be suspect. Knowing this, you can speak candidly to the sales people so that they can make favorable adjustments to your kit. If you are an informed consumer, you can build an incredibly powerful quiver for far less money than any great sale at *BestBuy*.

When I call these places, I approach the experience with the sales team from a position of apprehension – the person on the other side of the line is likely motivated to convince you to spend more money than you originally hoped. If you know this, your sales and service experience will be positive and save you a lot of money.

Final Thoughts on Quivers

My views about quivers are food for thought. I highly suggest that you carefully account for all of the gear that you will need and ask other people, who have filmed projects similar to those that you wish to undertake, about what equipment they use. When the opportunity presents, have them take you through you the pros and cons of their choices. If this is not an option, stop by the local mom-and-pop shop and ask the same questions. These folks usually have some well-informed ideas.

Chapter 7
Purchasing Postproduction Equipment: Video Editing Programs

Video Editing Programs

Lots of professional filmmakers will argue that one video editing platform (VEP) is better than most others. They likely have lots of evidence to back up those assertions. Indeed, these opinions are based upon years and years of experience. There is a good chance that their paths have been tumultuous, and opinions are tied to trying times and knowledge that blossomed from hard-earned breakthroughs. Technology and user interfaces have become more functional, strong opinions and scars persist.

VEPs are designed to aggregate and sequence media, including music, video, photos, narrations, and text, to tell the story that you want to unfold. Once these files are linked end to end, or overlapped so that they play simultaneously, your vision takes shape. VEPs offer you choices to manipulate the various media files to better serve your purposes. These edits range from very simple changes on up to very complex metamorphoses of the file's original manifestation. Just to wrap your brain around how complexity mounts, let's list some changes, from very simple and common to those that are more demanding.

- Shortening video files to remove unwanted content
- Turning down the volume on a musical piece so that you can hear a person's voice
- Adding transitions between separate video clips such that one fades into the other
- Adding text to complement a photo or video clip
- Animating the same text to shrink or grow
- Throwing a sound filter over a soundtrack to remove unwanted hiss
- Manipulating the color qualities of a video clip to brighten the scene and remove unwanted yellow light

These few examples are representative of how deeply you can delve into tuning how media files unfold and how they are linked together. You can accomplish many

© Springer International Publishing AG, part of Springer Nature 2018
R. Vachon, *Science Videos*, https://doi.org/10.1007/978-3-319-69512-9_7

of these tasks with a VEP that comes stock with your computer. Should you like more control over what you can do though, you will need to veer away from the free programs and step into those that give you more control over some of these edits. For example, careful color correction functions (dialing in the hues or contrast within different clips) come with upper end programming. However, learning complex editing programs takes time and familiarity. So, in the beginning be reasonable about what you can accomplish.

The greatest determinants for which program you might want are your motivation and video-making limitations. For example, what are your budget constraints; how many people you hope to collaborate with; where do you hope to be in 1 month, 1 year; or where do you hope your professional efforts will navigate? I posit that an decision, guided by answers to these questions, will get you closest to having your needs and hopes met. Once purchased, the greatest determinant of a VEP's functionality is the amount of time that you spend in the saddle using it. The bells and whistles are secondary. Place all of your efforts into learning your editing software and use its tools to accomplish as many of your needs as possible.

With time, you may learn that the program that you picked does not meet some of your most important editing needs. It is true that if you had known that you needed this function up front, you could have avoided using a VEP that you will replace. However, getting to learn another program is often not a monumental setback. Along your road of learning the first program, you likely practice options and methods that are invaluable for future efforts, even with another program. Why? The core gains that you make are intrinsic to editing, no matter the VEP – storytelling, pacing, melding media, patience, scene transitions, workarounds, and streamlined workflows. Learning the tools and shortcuts on a particular platform are essential, but those insights are rapidly learned in comparison to creating potent workflows and habits.

Inexpensive VEPs

Basic video editing is possible without expensive software. Most computers come outfitted with one VEP or another that gives filmmakers the tools to make simple edits and deliver a potent, albeit modestly spit-shined video. For example, Microsoft's *Movie Maker* or Apple's *iMovie* come stock with your computer or can be downloaded without cost. Incredibly, you can also download both for smartphones where you can edit complete movies while on the move.

These more simple VEPs teach you how to link media, such as photos, B-roll (illustrative shots – we talk about these in greater depth soon), interviews, sounds, and narrations, together into the storylines of your choice. For an individual hoping to make very simple videos, this is great; however, such VEPs present very few options for making refined edits to your films (e.g., correcting for color or poor sound). What's more, not all lower-end VEPs give you the flexibility of upgrading to more expensive editing programs while using a similar user interface. For example, Apple's *iMovie* (very good for intermediate editors) is very different from

Apple's more advanced editing program, *Final Cut Pro*. Unfortunately, if you do start a film in basic editing programs and later choose to edit with a more sophisticated program, you run the chance that some of your earlier editing labors will be lost. You may have to start from ground zero again – relocating your media assets and inserting them into place. This is time lost.

Note Simple editing can be done with online services that also help you with storyboarding.

More Powerful VEPs

If you are looking to invest in learning video editing and making compelling videos as a hobby or as part of your work, you gain quite a bit more power in manipulating your media by working on more complicated editing platforms. Some of the more popular high-end, prosumer VEPs include Apple's *Final Cut Pro*, Avid *Media Composer*, or Adobe *Premiere Pro*. These are at the top of the heap, but their user interfaces are simple enough to also serve as good platforms for those who want to grow their editing skills slowly. Complete with these VEPs are powerful color and sound filters, personalizable transitions, and titles with numerous intuitive tools for manipulation. A great example is image stabilization. An unstable hand on a camera can result in a terrible video clip. Ruined? Nope, with little time, image stabilization applications take some shake and vibration out of footage, sometimes turning the unusable into valuable contributions to your story. More importantly, you can dig under the hood of these options to personalize the effects. Believe me, you can go well beyond what you thought was possible. Indeed, with enough digging into effects, you might develop methods that have rarely been used to bring life to film. This means that a little up-front effort can get you started with using these VEPs; however you can spend a lifetime refining your knowledge and learning tricks for navigating options and making stylistic video.

When working with teams, it is advantageous to pass your video off to someone who can edit or refine, say, the sound quality of your video. Some VEPs are built with these needs in mind. As long as other users have access to the same raw video files as you, you can pass along the VEP files (associated with the video editing effort – like an Adobe *Premiere Pro* file that ends in ".prproj"), and they can then take over.

What's more, some of these programs give you the power to create gripping animations or digital renderings in a complementary program. By saving these files in the other program, the changes will be automatically added to your VEP, thus manifesting its changes in your video (without you having to import the new media). This is incredibly effective at streamlining workflows.

More complicated editing programs can run up additional costs. Above and beyond just the cost of the software, it is also advantageous to have a fast computer. Video files, particularly when filmed in high resolution or 4K, are cumbersome for editing and review, and these issues compound when you add digital effects (light and sound filters, or transitions) in your VEP. Whether on a PC or a Mac, intricate edits, large video formats, and complicated software perform better on a computer with lots of RAM and memory. As with any file or body of work that you value, backing up the files on a secondary hard drive (or using Internet services) is very important. This mounts even more expenses. Quality does come at a cost.

Final Thoughts on VEPs

VEPs can be simple or complex. You can get lost for days in figuring out how to conduct precise edits. Is this what you want? Knowing how far you want to take video editing is critical to VEP shopping. If you are on the fence about whether making films is what you want to invest in, start small and see what you think of video editing. Most people figure out whether it is for them or not quite quickly. If you are a beginner, yet have goals of grandeur, the road to complex edits is long but very satisfying. Set a pace of learning that you can handle. A little consistent effort pays more dividends than jumping out of the gate hoping that you will learn everything right away.

Chapter 8
Preparing for Filming

Practice with Your Quiver

The video assets that you collect are the foundation of your video. Poorly filmed video clips limit the utility of your film from the very beginning. For example, well-filmed clips are appropriate for more places within a film. In some cases you could go back and refilm a scene. Or, once back in the editing studio, you can manipulate some suboptimal video clips to suite your needs. Or you could go back and rework your storyboard to accommodate slightly deficient video content. All of these stopgaps are unnecessary wastes of time and resources if you film for success the first time around.

Familiarizing yourself with the equipment well before important film days sets you up for getting the best footage possible on the first try. Be mindful that throwing yourself into the fire of recording mission-critical clips the day after your quiver comes in the mail often results in mediocre footage or worse. Instead, devote unstructured time to having fun with your new equipment without the pressure of performing. For example, put yourself into difficult lighting scenarios and figure out how you can adjust light settings on your camera to optimize your shot. Try to tackle any shot that you can imagine, even the tough ones. That way, when you are confronted with challenging scenarios further down the road, you are confident that you can overcome them.

I have benefited from going out on filming outings with friends who have camera experience or similar interests in growing their skills. Learning these people's perspectives about what they hope to capture and the framing that will best make that happen, is mind expanding. Consider working with a friend or collaborator capture a shot that represents a theme, like autumn leaves. Each of your sensibilities and skill sets will deliver different results. Appreciating their logic will increase your versatility and expertise. Even if you film on your own, share your best clips with your friends – tell them what you were hoping to accomplish, and ask them what they think and how you could improve them (Image 8.1).

© Springer International Publishing AG, part of Springer Nature 2018   101
R. Vachon, *Science Videos*, https://doi.org/10.1007/978-3-319-69512-9_8

Image 8.1 One can learn a lot about how to use videography to convey content and tone from others. Join a friend for an afternoon of filming. Compare how each approaches specific filming challenges. You might try to "capture the environment." You'd be surprised about how similar themes can be communicated so differently

In the beginning, you are on the steepest learning curve of your film-making career (or whatever you think that it is best called). Brutal honesty can hurt, but it is better to learn early on in the process then when you are rushing to release a finalized film. The more open you are to new concepts and failure in the beginning, the more you will lock in the good habits, other's opinions, and time-saving methods.

Gripping footage takes a little more time and effort to set up. It's not just having the camera running that creates these opportunities. The filming of well-composed shots with the resources at your disposal increases the potential of your work.

Preparing to Get Out the Door

A filming day begins before your step out the door. Being prepared relaxes you when you are in the heat of a difficult shoot, and ensure that you are going to set up and take the right shots to fulfill your storytelling needs. Here are a few simple preparation methods that I have found useful.

Thoughts for Success

On filming days, it is likely that you will have countless things on your mind. Schedules, lighting, keeping people in position, monitoring sound, or any host of other tasks add distractions or lead to being overwhelmed. Just the same, you are there to film quality shots, and how you approach filming is indelibly recorded in each shot. So, groundwork includes organization and bringing the right attitude. For starters, learn the self-talk that works to excite you for a long period of hard work. All of us have ways to coach ourselves into getting hard-won tasks accomplished. What do you need to hear or understand to motivate you to put 15% more effort into your work? Draw upon this language as you prepare for a day of filming.

Personal Anecdote
I am a climber. I scale high rocks and icy cliffs using handholds (cracks or ledges) that challenge my body and mind. I swing ice axes into vertical curtains of ice. I am drawn to the process of solving the ways that my body can be positioned so that I can reach from one hold to the next, sequentially (Image 8.2).

Sometimes the goal is to climb fun, easy routes. Yet other times, I deliberately go to climbs to use everything that I have (power, agility, my mind, happiness) to make it to the top of a very difficult climb. These efforts strain tendons, hurt the skin on my fingertips, and expose weaknesses in my logic. This is literally painful, and failure to succeed can feel like a hit to the ego. If I enter an attempt to a climb route, without my highest level of enthusiasm and positive energy challenging, I usually won't succeed. In order for me to climb harder and harder routes, I have had to learn how to self-talk my way through times of doubt. Negative talk can spiral into giving up when I am close to success. Negative talk means that success is deferred to another day. Of course another day will dawn and there is a chance that I could return to the same rock; however, maybe success was possible on the first day?

Looking back over my successes with climbing, I have found a handful of very simple exercises to focus my mind on the task that I am undertaking. Sometimes, I ask myself, "Do I want to come back to this same effort another day, or do I want to do it today?" or assert "You are strong enough, so don't let your mind or mood keep you from success. If you fail, it is because you let failure seep in. Tomorrow your mood will be different, but the outcome will not change." Or, "in a few short minutes you will be done, success or failure, you will be done. How do you want these few minutes of pain and confusion to manifest? What must you release for a limited amount of time to make success a viable option?" Alternatively, there are little habits that I try to maintain, such as eating and hydrating. A low-energy feeling infuses doubt. Or, when I am confronted by a move that confounds me and I fear falling (I have a rope attached to me; however falling into air can still be scary), I smile. Yes, the simple act of forcing a slight smile is linked to changing the attitude of how I approach a problem.

These are words that have helped me win world-class climbing competitions and finish rarely repeated routes. They are proven methods to create the results that I seek – they raise me above the mire that drudgery or pain present.

Image 8.2 The author ice climbing

Let's connect these thoughts to discussions about filming.

You, the videographer, have to be dedicated to telling your story. You can make small mistakes, but recovery and completing your filming hour, day, or week with style and buoyancy is very important. If you are new to using filming to tell a story, this notion might seem a little dramatized. For informal filming excursions, that is probably true; however, try a 10-hour day. Long filming sessions test even the most experienced videographers. Negative self-talk becomes evident when you go back to review your footage. Once in the editing studio, you can wish that you put in a little more effort to move a light or monitor background sound. Identify and practice some thoughts that ready you to bring your A-game to filming.

Create a Gear Checklist

It is awful to show up to a shoot and not have a spare battery or memory card. The entire filming session can be shut down early. Batteries are just an example of any number of pieces of gear that, if forgotten, can set back an effort. Avoid this happening by creating a gear checklist. My list also includes what I need to keep my energy level high and produce success. The length and items of these lists range with different applications. Before you head out the door, go over what you have and where you keep it in your camera bags. Here is an example checklist that I might carry into the mountains for a landscape shoot.

Essentials	Supplemental
Camera bag	2 Snicker's bars
Camera	Headlamp
Battery 1	Duct or gaffer tape
Battery 2	Water bottle
Battery 3	Survival kit
Memory card 1	Sunscreen
Memory card 2	Warm layer
Memory card 3	Handkerchief
Tripod, tripod head	Baseball or stocking cap
Shotgun microphone	Notepad and pen
Short slider	

Prepare a Shot List

Once on-task, remembering the shots you must capture can be difficult. Maybe the night before, review your storyboard and write a checklist of the shots that you know you need. Additionally, I have found that a few minutes of in-advance, creative thinking helps me identify how I can make shots come to life – taking clips from good to great. This exercise lends structure to my filming day and, as a result, reduces the chances that I will be hitting my forehead in frustration a day later because I didn't remember to film very specific clips. Even if I don't refer to this list as I run from one place to another filming, the prior exercise might trigger memories about what is needed. Plus, it returns me to the mental place – understanding that different scenes will carry your viewer through the story. To be missing a clip could blow this continuity.

Perhaps you are going to capture an interview. The list of shots might seem limited to those of your interviewee but step back for a moment. What topics are they going to be talking about and can you take? Can you take descriptive of what you are hoping to discuss? How about taking a setting shot where the interview occurred? Or, are there additional people that you would like to interview while on location? There are always more shots that deserve attention.

If you are anything like me, I might forget to bring something that I typed on my computer. Printing out a copy of your shot list gives you the option to physically check off all of the shots that you wanted to capture. Plus, paper is often easier to transport and open while running and gunning. What's more, if there is a chance that it will rain during your shot, it is worth getting a piece of paper wet over your computer. So press print and don't forget to grab the paper from the printout tray!

Day of Filming: Getting It Done

Let's think about what you need to bring or do on a filming day in order to increase your chances of success. Some of the hints that I discuss below describe physical tools that can up your filming game, while others are mentalities that I have found to help my filming ventures.

Something That All Videographers Must Know About: B-Roll

A part of being prepared is to keep a sharp eye for good *B-roll*. *B-roll* is footage that illuminates concepts, so that viewers can see what an interviewee or narrator is talking about. If a narrator is talking about a dam, the filmmaker will show imagery of a dam. What's more, B-roll gives the editor the chance to cut and splice an interview to be more meaningful and thus useful. What do I mean? Let's say that your interviewee says something very useful for your final video, however the good explanation dissolves into a minute of free-flow conversation about something tangential to the needs of your film. This can be frustrating to see as an editor. "Why didn't they just keep their comment tighter?" Well, they didn't but you can trim their words for impact later in your editing studio. B-roll help to hide any cuts that you might make to these scenes flow more effectively (Image 8.3).

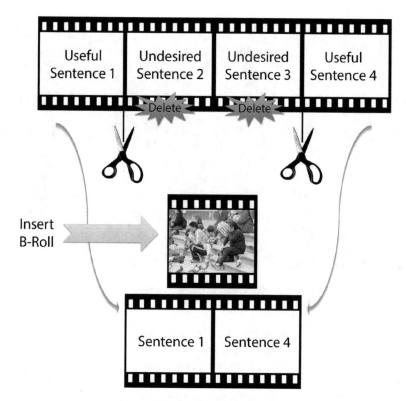

Image 8.3 B-roll is a useful tool for when you want to retain a stretch of sound from a sequence of edited video clips. Sound may play fantastically; however, the video jumps in framing. This can be covered up. You can trim out undesired sentences and cover the jump between sentence 1 and sentence 4 with some illustrative footage

B-roll is one of the bread-and-butter tools for a filmmaker. As such, it is critical that a videographer/editor keeps their eyes peeled for supplemental shots that can best illustrate their points. The more B-roll shots we have, the more choices we have for building a strong film. If you are the storyteller of your film and you are the one to procure media, from the conception of the film start scouring and archiving useful B-roll shots, articles, or animations in a place where you can readily find them. Here are a few questions that might guide you towards collecting this B-roll footage, photos, or figures for your next film.

What shots best illustrate a process? Start taking video and pictures of anything that might serve as equivalences. Search personal photos or those online. Save them in folders on your hard drive that are easily searchable. The worst-case scenario is that you never use the shots. The best-case scenario? A little effort makes your video shine! You might be surprised about how well this approach works to give you options in the editing studio.

If your work revolves around an abstract concept, metaphors work wonders to clarify what you are trying to explain and are tools to increase further engagement of your audience. For example, perhaps chaotic thoughts could be shown as a flock

Image 8.4 Videos metaphors are great ways to convey hard-to-illustrate concepts. Here, a hoard of teaming birds could represent poorly directed thoughts or a traffic jam

of seagulls bobbing and weaving? Or a monstrous mountain could represent the monumental need for funding (Image 8.4).

Additionally, consider what shots best illustrate outcomes. Record video of your technical deliverables in action (data, device, or community health campaigns). If your software translates data into the most engaging, colorful, and geometric graphs, take a screenshot of your most prized samples. If you worked on an inexpensive way to draw water from wells, and as a result have had huge impacts on communities in the developing world, build an archive of media taken during trips into field. As you will shortly learn, even if you are only able to capture still shots, there is great utility for such media in films!

Hit the Core, then Explore

During a filming session we are often confronted by clashing needs, communication gaps, problems with technology, and complications with settings. Appropriately, it is not unusual for a day of filming to stretch on quite a bit longer than originally planned. As such, prioritize how you will capture your footage. Whether you are filming wildlife, laboratory experiments, or interviews, secure the shots that will guarantee a successful day of filming. If a day is broken down into filming sequential scenes, with several takes each, save the most innovative and difficult shots for the last attempt in each scene…and then move to the next scene, and repeat. If you get trapped into being incredibly creative, you might come away with an incomplete catalog of clips. I like to call this, *hit the core and then explore* - A simple guiding mantra to every filming day.

Film with Editing in Mind

Much like the purpose of your shot list, film with your final video in mind. Sure, you can capture shots that follow the equation outlined in your storyboard; however, filming is an active process that entails subjectivity, an aesthetic eye, and attention to the voice inside you that says "the last shot was good, but this shot will win out." The former is a sensibility that comes with experience, but it is healthy to remind yourself that while you are filming, staying in touch with the purpose of a certain footage will guide the way that you compose a scene. Listen to these inclinations.

Redundancy

Mistakes happen while filming, transferring files, or later on in the editing process. Files can be corrupted or your sound quality can be compromised. What's more, wildlife or people are fickle, and one shot may skimp on the chance that a second shot will be better. This notion expands into scripted scenes – a second or third (or tenth) "take" (or version of a scene) may come off better than the first. You might not think that you can do this for interviews, but on occasion this is acceptable. A simple way to get someone to answer a similar question for a second time – ask them to phrase their thoughts a little differently, perhaps telling them to condense or expand their thoughts. Most people are very open to condensing or rephrasing with a slightly different emphasis. As such, take redundant shots, thereby reducing the chances of running into roadblocks caused by errors or increasing your chances of taking a scene from good to great.

115% Mentality

Today is your day to make a difference with your film. Your final film will benefit from you taking all of the appropriate steps to capture the right shot. Don't settle for less because you did not enter filming with your best efforts. What is one way to envision this? Treat filming like parenting. Emphasize vigilance to the thoughts and methods that consistently make for excellent footage. Is my audio good? Is the tripod stable? Does my interviewee have one collar up and the other down? Is light glaring off of a windshield? If it helps, turn those questions into a chant! OR, take the extra 10 s to assure that your shot will capture what you need. Stay an hour longer on location to be there for the sunset or the finish of a race.

Just the same, give yourself over to patience with anyone that you are working with. Think about how you can assert needs, confidence, and understanding of what you need to accomplish, without becoming a tyrant. If you are hungry, take a moment to eat. This will give you lasting performance with your camera and

Image 8.5 Effective filming has much to do with interactions with other people. Filming and future opportunities mount when you go out of your way to build strong relationships, help, and maintain a positive attitude

patience with collaborators. As an investment in long-term vigilance, make sure to have water and food on hand while you film. Not just for you! If others are integral to success, have spare snacks for them. Not only does this make them happy in the moment, partnering individuals will look back on the experience more fondly, potentially opening doors for future ideas. At the very least, they might offer a fond pat on the back when you see them at a coffee shop or conference the following year (Image 8.5).

Concluding Thoughts on Preparing for Filming

You can almost be assured that every filming day will present its own challenges. When on a shoot, small glitches can place blinders on creativity and reduce the potency of your efforts. A major step to averting these issues is to identify which ones could present, and then taking up-front action to reduce their impacts. Building in time to prepare is a great habit to begin early on in your filmmaking career.

Chapter 9
Stationary Filming Techniques

Stabilization

As we discussed in Chap. 6, when we introduced the tripod, it cannot be overstated that most shots improve with a stabilized camera. You might be tempted to quickly shoot with the camera in your hand; however, reconsider that choice. Viewers can absorb a landscape, the words of an interview, or the methods of manufacturing a device when they are not distracted by bobbles or shakes. As such, tripod use is recommended.

Techniques for using a tripod, which can simplify the process of filming and take shots from good to great, require knowledge and skill. Typically, this means that all of the tripod legs are in contact with flat and solid ground. Additionally, the head of the tripod should be centered over the legs.

Squaring up a tripod in unstable or uneven terrain can present challenges. Early in my days of video, I remember kicking myself over poorly positioned tripods readjusting while I was executing a pan. Imagine a half-hour time lapse of clouds and the camera repositions or vibrates. These foibles are devastating. Thoroughness with tripod positioning is now engrained into my filming practices.

What about my own positioning as I film with a tripod? I place one leg slightly behind the other, with feet shoulder width apart. It is tempting to keep my hand on the camera; however, I warn against it. Particularly light cameras coupled with inexpensive tripods are prone to slight movements that all hands make (no matter how steady we hope to be). The same is true if you move your face too near a viewfinder on the back of the camera – slight bumps diminish the value of a shot. Unless in high winds, press record and step away from the camera, content to watch the recording from an arm's length away (Image 9.1).

The only time that the camera should be positioned away from the center of a tripod is when you seek a unique camera angle where either the framing of your shot or the ground can't accommodate the ideal position. When might these rare occasions happen? If you wish to film steeply down towards the ground, as though you are filming insects, or up towards the sky to capture a plane. In some of these cases,

© Springer International Publishing AG, part of Springer Nature 2018   111
R. Vachon, *Science Videos*, https://doi.org/10.1007/978-3-319-69512-9_9

Image 9.1 Unless you are going for a specific feel, it is a great habit to practice filming with a tripod. Level your camera over equally weighted legs of your tripod. Take a comfortable stance a couple of feet from your camera, with feet shoulder width apart. After pressing record, keep your hands and face from touching the camera until the shot is finished

Cool Trick #1

Let's say that you are running and gunning. Every second counts and you have a million shots to take. Perhaps the majority of your shots are to serve as B-roll, and you are pressed to capture all of the shots that you need. The first step is to continue using your tripod, but for each shot that you take, change the framing for a second or third shot (by either zooming in or zooming out). This way one tripod setting and singular aiming of the camera will gain you additional useful clips. While you record, look about you and determine if you can quickly steer your camera to a new angle and take even more shots. Sometimes I have recorded half a dozen or more unique clips without moving the legs of my tripod.

the range of pivot for the tripod head is not enough to film in these directions, so you have to tilt the tripod over two legs. Another occasion might be if you can't find firm placement for one of the tripod legs, and the second best choice is to use only two legs in conjunction with a steady hand (Image 9.2).

Image 9.2 Uneven terrains may dictate that you use modified techniques in order to film stable shots. If the ground is too rough for three stable legs, use two, then rely on your stable body to act in place of the third leg

Cool Trick #2

Still shots are helpful media for producing films. Photos that DSLRs and many camcorders snap are another viable solution. They create a stable video content in situations or conditions that are not ideal for tripods. How? Most often, the resolution of a photo far surpasses that of a high-resolution video. As a result, still shot photos may do a better job at catching the framing and crispness that you desire from a video shot. What's more, if the objects that you hope to film move very little or stabilizing your camera is hard, photos can cure these problems.

Once back in your editing studio you can crop in on photos (Image 9.3). Using the increased resolution of photos also affords you the chance to show activity in photos. How? Digitally manufactured zooms and pans within your video editing program (VEP). Unfortunately, foreground and background elements in a scene relate to each other differently when moving through a photograph rather than when a similar shot is captured with moving video. However, leveraging photographs to augment a collection of B-roll is a common practice, and have bailed a lot of professional filmmakers out of considerable production binds or content to support a compelling story.

Image 9.3 Most video cameras give you the option to take photographs in addition to filming video. Resulting still images are often several factors higher in resolution than video. This higher quality gives you some leeway when filming in challenging situations. Photos might give you more options later in the editing studio than video because they are more stable. What's more, you can crop into a photo more easily, thus drawing a viewer's eyes more effectively onto core elements within a scene

What happens if you don't have a tripod on hand to steady your camera? Because image stabilization is still a priority, sometimes leaning your body and your camera against a tree, rock, or post will suffice (Image 9.4). If that is not an option, you have to depend upon yourself to stabilize your camera. Some camcorders position upon a shoulder, thereby distributing the weight through your body. These cameras are quite expensive, and not very easy to wield and transport when wanting to move efficiently. Other camcorders are designed to comfortably hold in front of your face. Even though this position is less stable, you can work with this. Hold your arms in close to your chest. The key is to create a fixed arrangement between your camera and your shoulders, chest, and/or hips. Using a connection to the weight of your body's core smoothens out some of the higher-frequency shakes and bobbles. What's more, to a certain limit, the heavier the camera, the fewer the high-frequency shakes that they pick up.

Note Tripods are a great place to begin when filming. With experience you can learn equally effective methods for filming that incorporate a monopod (a single leg stabilizing device) or your body.

Image 9.4 The quality of video improves if you find ways to stabilize your camera. Without a tripod, lean on creative solutions to remove wiggles and shakes from your shots

Composition

Composition includes all of the elements, and their positioning and movement, in a scene that communicate information and give a shot the feeling that you are going for. Framing of a shot is critical to crafting a shot's composition. Framing defines the boundaries of an image and how objects within the scene relate to each other spatially. In general, the framing of scenes can be broken down into wide, medium, close-up, and extreme close-up shots.

Wide Shots

I like to think that *wide shots* are ideal scenes for showing the spatial coverage of a specific activity. They give context to the story. Perhaps it is a shot of a cityscape, an office building, or the activity within a coffee shop. I think that wide-angle shots are perfect for defining setting and tone, without the need for character involvement; however, if a character is involved, it shows them integrated into their environment – perhaps ordering coffee or walking down the street. I like to use these kinds of shots to start a story, even if I am trying to describe a science concept. For example, the loneliness of a laboratory at night. The charged energy of a packed concert stadium. The calm of rain falling on a lake. Later, you can zoom in on the details of the story, and the viewer knows exactly where the story is happening (Image 9.5).

Image 9.5 Two examples of wide-angle shots

Medium Shots

A *medium shot* shows a subset of the subject matter in a scene. The process of capturing this subselection can be accomplished either by physically moving your camera closer to the setting (compared to a wide shot) or by zooming in with your lens. Some videographers suggest that a medium shot strikes a balance between the details of the elements that tell your story and the setting that holds these elements. It gives enough detail to see where the action fits into the broader setting while

Image 9.6 Two examples of medium-angle shots

delivering hints of the finer details of your story. The framing of an interview would include an individual's waist to the top of their head. Perhaps it is a football huddle, if your wide shot is of the field. Maybe it is five people around a conference table, when your wide angle was of the entire building. How about an operating table with a patient and the technicians? (Image 9.6).

Close-Ups

Close-ups are exactly what they sound like. You direct your camera in close to the most important information. All other distracting information falls away and the viewer is drawn into a microcosm. Close-ups are effective at focusing the audience on details. Details can be methods, how a device functions, or emotion-filled facial expressions. For example, a close-up may capture only an interviewee's face and shoulders. The viewer is drawn into expressions and mannerisms of the individual and cannot avoid seeing the human behind the words. Close-ups scream to the viewer, "pay attention!" Note: if you film too many close-ups of unimportant elements to your story, people stop paying attention to the greatest value of close-ups: importance! (Image 9.7)

Image 9.7 Two examples of close-up shots

Extreme Close-Ups

Extreme close-ups draw the viewer even further into the detail of the scene. They are incredibly powerful ways to highlight the importance of something small. A bead of sweat on an athlete's nose. The singularity of someone's finger pressing the red button to drop a bomb. Dust blowing across the floor of an abandoned warehouse. Extreme close-ups are painstaking shots to prepare, but serve as explosive tools to knock home intensity. That said, they can rarely carry a story on their own (Image 9.8).

We brought up some examples of wide, medium, close-up, and extreme close-up framing. I am sure that images come to mind that fit into these categories. However, how we describe these framings is largely dependent upon the context of each scene. For a story about ants, a wide shot might simply embody a stump, the medium shot the runnels of bark, and the close-up captures a string of ants carrying leaves in a line. The extreme close-up might show mandibles tightly gripping a stem of a leaf. Alternatively, a discussion about the structure of our universe may use an animation that passes

Image 9.8 Two examples of extreme close-up shots

numerous galaxies as a wide shot and the medium shot illustrates individual solar systems. The close-up might be a singular planet and extreme close-up canyons carved by theoretical rivers that once existed on this planet. When defining what is wide or close for your shots, knowing the information you wish to convey is everything.

Regardless of the content of your story, a healthy variety of shots is helpful to telling complex stories and are useful for retaining the attention of audiences. Transitioning from wide, middle, and close-up shots is an art. When wielded effectively, changing from one framing to another is immensely useful for carrying a story and strengthening the tone. Unfortunately, there is no hard-and-fast equation for how often you should transition between scenes. Knowing how framing affects a viewer or bolsters story elements can help you define your use of framing.

Cool Trick

Well, this is not really a cool trick so much as a helpful mind-set as you set up shots to serve different framings. Avoid empty spaces in your shots. Your job is to use the objects and characters in your scene to support the feel that you are striving for. Rarely are empty spaces, also called *dead spaces*, useful. Take a look at the following shots and see how the dead spaces pop out and how they leave you feeling. Most viewers speak of dead spaces as though they feel left hanging or are experiencing an incomplete thought. Something surely must be missing. People's attention is drawn to curiously empty spaces and stray away from storylines. If an object moves to fill this dead space, this can be useful; however, if unused they present distractions. As you frame a shot, accounting for the movement of elements, try to avoid these spaces (Image 9.9).

Positioning for a Shot: Moving in Versus Zooming in

How do you get to the right position for camera framings? For voyages through the galaxy, you will likely leverage your imagination to create an animated shot. If you are shooting your scenes with a camera and lenses, it is up to you to interact with your environment in a way that best stages the shot that you seek. This is not easy, and extra work pays dividends on potent final video products (Image 9.10).

One of the most dynamic, yet flexible, ways to adjust the framing of a shot is to pick up your camera and move it to the precise location where the angles and appearance are just right. Constantly moving your camera is time consuming, but delivers on specific and compelling feelings to your footage.

Opposite to transporting your camera to different positions in relation to the elements in a scene, you can zoom in or out with your lenses and capture your desired framing. However, the relationships between elements within a shot are different for zoom shots in comparison to those that you physically move towards. Zooming gives the shot greater interactivity between moving foreground and background elements. Small foreground movement is linked to large relational shifts with

Image 9.9 (**a**) Dead spaces in a shot distract from useful details or an unfolding plot. Here, we hoped to use a keynote presentation to share information showing beside the presenter; however, because of lighting, the screen over exposed and goal was not met. Later, we filled in the space beside the presenter with still shots from her presentation (Consortium for Capacity Building). (**b**) Sometimes expansive areas of emptiness within a shot add to a feeling of openness or journey. In the case of our second shot, the hiker moves through the openness, fulfilling the need for him to move towards a destination

Image 9.10 In most situations you can set up shots without too much thought. However, filming from unusual locations and perspectives can serve as potent tools for conveying setting. Go out of your way to safely capture the shots that will further grab an audience

background elements. What's more, the focal length between objects may shorten as you zoom in. For example, you may focus on changing leaves on a branch, and the rest of the forest will be out of focus. This is a powerful mechanism for drawing a viewer's eye to specific objects and ignoring others (Image 9.11).

As you position your camera in closer (rather than zooming in), the foreground appears larger and background elements smaller. From one perspective, foreground elements might take on greater meaning because they are proportionally larger than other elements. This also means that the dimensionality of foreground objects (such as an interviewee's nose or forehead) may appear exaggerated. This creates a stylistic appearance that may work either for or against what you are trying to convey. On one hand, filming conversations up close feels personal and intimate, while zooming in from afar lends a feeling of distance and authority of the speaker. Why? The viewer automatically assigns a feeling of distance and removal from someone if a telephoto lens is in use. This locks in the notion that shots captured by telephoto and wider-angle lenses evoke very different feelings. If you are filming a bolt attached to an engine, this may matter little in your storytelling; however, the use of the right lens when filming facial expressions can greatly strengthen your story (Image 9.12).

Wide shots are often filmed with wide-angle lenses, whereby the camera lens allows a videographer to be, say, in the same room as the subject matter, yet still nab all the desired subject matter. Lenses set on wide-angles capture lines within a shot that give the viewer a feeling of being close to a scene, but seeing everything. One

Image 9.11 Zooming in, when coupled with a short focal length, is a powerful tool for drawing out specific information in a scene

can still film wide shots with a lens set on zoom; however, to do so, the camera must be positioned quite far from the subject matter. What's more, wide shots captured by a zoom lens create lineage within a shot that leaves a viewer feeling like they are far from the action. Usually these far away shots take more effort, so videographers often move closer and film with wider-angle lenses.

Working Layers and Angles

Use layers to your favor. Once you understand that you, the videographer, can define how elements in the foreground and the background interrelate, you can create a more engaging spatial composition. Don't hesitate to include some angles that exacerbate these relationships. Intersecting diagonal lines tell a

Image 9.12 One has a couple of choices when executing image framing – either zooming in with your lens or physically moving your camera to capture the desired frame. However both choices treat the geometry of elements within a scene differently, thus evoking different results with audiences. Notice how the positioning of background elements differs between the first, zoom-in shot, versus our second, move-in shot. Some say that with interviews, moving in gives a more personal feel, while zooming in is more professional

viewer how different objects are trending or relate. If the opportunity arises, use light to enhance a viewer's understanding of where each element fits into the setting (Image 9.13).

Speaking of angles, framing can also allude to the aspect (or direction) at which you point your camera towards a subject or setting. Changing angles modifies the way your viewer looks at familiar objects. Different perspectives add dimensionality to a story, give the viewer a greater understanding of setting, and can serve as a fantastic way to break monotony of fewer shots. For example, when looking at a singular object of interest, like a car, I generally try to change the angle by no less than 30°. This is known as the *30-degree rule*. Anything less is too oddly similar to the previous shot and thus becomes a distraction.

Image 9.13 Angles and layers add dimensionality. Try to position your camera such that the lines of the foreground and background interact, making for more creative and impactful shots

Summary of Compositional Pearls of Wisdom

Becoming comfortable with composition takes time. I find it helpful to keep in mind a few pearls of wisdom that deliver on more potent shots. Some of these ideas have come out in earlier discussions and a few more are worth mentioning.

- Level your camera to the horizon. Tilted shots are immediately distracting. Inexpensive tripods often come equipped with heads that are challenging to level accurately. Consider changing the length of different legs to help serve that purpose.
- Take extra care to place the core elements in a scene in good lighting.

- Wide-angle shots are fantastic at establishing settings.
- Medium shots convey interactivity and feeling simultaneously.
- Close-up shots draw out a person's emotions or embed the viewer deeply within a process.
- Wide-angle shots establish setting, yet closer-up shots are more effective at conveying information. There is nothing more dislocating than using a shot that pivots on details, however the details are too small to see. A viewer can feel left out of the story.
- To build upon the previous statement, don't be positioned too far away from your subject, particularly when conducting an interview. Personality is lost.
- Transitioning between framings adds depth to a story. Too much of one framing gets boring and your content may lose impact.
- Backgrounds can be distracting. Sometimes too much movement or detail in the background draws attention away from the focal points of a scene. Two quick solutions: seek out less complicating backgrounds and/or change the focal length of your lens to blur out the background (we discuss focal lengths in greater detail shortly).
- Scenes varying only by slight changes in framing tend to frustrate viewers.
- If you are looking to keep your camera on wide, medium, or close-up shots, think about changing the angle of the camera by 30+ degrees between shots. When contiguous scenes do this, also called *cut angle*, they enhance viewer engagement.
- Viewers are quick to pick up on, and grow tired of, regularly spaced cuts. Avoid the use of clips that all last, say, 10 s. Diversity in clip length keeps audiences engaged.

Chapter 10
Lighting

How Your Camera Adjusts to Compensate for Lighting

The automatic light settings on your camera are a great first approximation for how light should be represented in a scene. However, under certain conditions you can depend on your camera to reliably miss the mark. Unfortunately, your camera does not know your needs. Keeping your needs in mind and making thoughtful changes to camera settings will optimize your shots. Many videographers find it exciting and empowering to know how to let go of the reins of automatic settings to make manual adjustments. Some assert that commanding the light settings on your camera is one of the largest steps that you can take to change your filmmaking from a very structured exercise tool into a very personal adventure in art.

Taking on the challenge of making manual changes to your camera demands that you evaluate the character of your shots, say landscape, sports, and portrait, and make appropriate adjustments to create a stylized feel. If you are interested in using these options, I would suggest getting outside and practicing methods long before an important filming day is scheduled. With enough experience, you too may reject the use of any automatic settings-trusting that your brain is much more skilled at making adjustments than any circuitry.

Camera Controls

Let's discuss how cameras can be adjusted or self-adjust to account for variable lighting. What's more, in order to tackle some of the challenges associated with variable lighting with onboard adjustments to your camera, we must delve deeply into the mechanisms within a camera that darken or lighten a shot. Even more, we will discuss how manipulating shutter speed, ISO and aperture settings create certain results.

© Springer International Publishing AG, part of Springer Nature 2018   127
R. Vachon, *Science Videos*, https://doi.org/10.1007/978-3-319-69512-9_10

Shutter Speed

Shutter speed refers to how much time a camera exposes the image sensor to light during every frame. *Note: Image sensors are the part of your camera that translates the colors coming into your camera at different locations into digital information that reproduce these patterns into a photo or video clip.* For cameras set to record at 30 frames per second (fps), it will capture 30 separate frames (images) within a second. As such, it makes sense that the shutter speed on many video cameras cannot go any slower than 1/30th of a second – 1/20th of a second would mean that the camera could capture a maximum of 20 frames per second, which is not a standard frame rate for video (remember standard frame rates are 24 or 30 fps). That said, shutter speeds can actuate much faster than 1/30th of a second. In these instances, a video camera rapidly captures images, say 1/250th of a second, every 1/30th of a second. This way, 30 images will result at the end of a second of filming.

Why would a camera change the amount of time that its image sensor is exposed to light? The longer that the sensor is exposed, the more light that it receives, thus a shot is brightened. Appropriately, long shutter speeds are preferred for lower light settings. For highly lit settings, videographers often choose to raise the shutter speed to reduce the amount of light hitting the image sensor.

Shutter speeds not only affect the brightness of a video shot, but a practiced eye can perceive other side effects. How do these manifest? Particularly in high-action shots, clips filmed at slow shutter speeds, like 1/30th of a second, have a blurred appearance, while any shot filmed above 1/250th of a second will appear more crisp. As such, if you start to monkey with shutter speeds on your camera and experiment with appearance, you might be one of the few people who are very sensitive to these subtle differences.

Be warned, if you are looking to take photos at the same time as video, slow shutter speeds require a very steady hand, or use of a tripod. In general, I would hesitate to take a photograph at shutter speeds lower than 1/30th of a second (when at a wide angle) and 1/200th of a second (when zoomed in). Why is this the case? If you calculate the geometry, a small shake in a camera when the lens is set to a wide angle will do very little to move the objects in your field of view. Oppositely, a small movement when zoomed in can make for enormous instability in a shot, thus large movement in objects around your image framing. Large movements in subject matter over a short period of time will result in blur (Image 10.1).

Aperture (Also Called f-Stop)

A camera's aperture serves the same function as the iris of your eye. The aperture controls the amount of light passing through a lens and on to your image sensor (over a given amount of time). The more open the aperture, the more light to hit the image sensor. The smaller the aperture, the less light hits the image sensor. The measure of how open or closed your aperture is called f-stop. The larger the f-stop, the smaller the aperture. Under high-light scenarios, automatic settings usually react

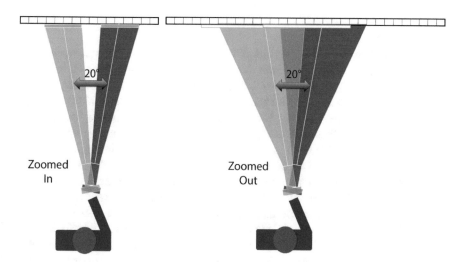

Image 10.1 Zooming in on a scene can better focus a viewer's attention on certain elements; however, zooming in also exacerbates any instabilities of your camera. Here we dissect how a 20-degree shift in the direction of your camera can leave most of the elements within a wide-angle scene framed, while every element in an original framing can be lost when zoomed in. As such, be careful when zooming to stabilize your camera. Little shakes add up to huge camera movements. Alternatively, if you can't stabilize your camera, set your lens to a wide framing and step in closer, thereby getting close to your subject matter, but reducing the impacts of instability

by slowing the shutter speed or closing down the aperture. You can make similar decisions when using your camera on more manual modes.

Videographers often adjust aperture and shutter speed in unison. Let's say that you wish to raise the shutter speed of your camera to film a high action shot. Opening your aperture will compensate for the lost light as you increase shutter speeds. The same is true for hopes to decrease the size of the aperture. In order to close your aperture, consider slowing your shutter speed.

This last comment does present the question – why would you want a closed aperture over one that is wide open? Adjusting your aperture affects more than the amount of light hitting your image sensor. It also changes the depth of field within your shot. What does *depth of field* mean? It denotes the ability of your camera to focus on both elements in the foreground and background simultaneously. Literally, the depth is the distance between a foreground object and one in the background that are both in focus. A shallow depth of field means that crisp focus can be accomplished for mostly foreground or background objects, but not both. A large depth of field means that things close up and far away are in focus at the same.

For many, depth of field is a useful tool. I find great relief when I am running-and-gunning to depend upon a great depth of field. The increased range of focus means that I do not have to pay as much attention to precise focusing, which amounts to time saved. If I have my aperture closed all of the way down (which equates to a high f-stop) I can move between shots with more confidence. There are other interesting effects that result from tinkering with your aperture that we will discuss a little further along in the chapter.

ISO

It is impossible to have a conversation about making camera adjustments to compensate for light conditions without mentioning ISO. Its origin comes from earlier days when photography relied on actual film. It referred to film's sensitivity to light. Low numbers means that the film needs more time to bask in the light from a scene in order to manifest the right level of brightness. In return for the time, low ISO film (about 50 to 200) delivered amazing colors and crispness. As you go up higher in ISO numbers, film needs less time exposed to the light coming through the lens. As such, in order to capture a photograph, your shutter speed could increase, thus reducing adverse side effects such as motion blur. Unfortunately, as the ISO number increases, so does the granularity of a shot. A photograph resulting from ISO numbers over 800 has reductions in image quality that some photographers notice.

For digital cameras, the term ISO refers to an image sensor's sensitivity to light. Uniquely, users of digital cameras can make instantaneous adjustments to the ISO, whereas with film you have to shoot an entire roll of film set at a particular ISO number before you could make a change. With digital cameras you can step from a dark laboratory to a brightly lit outdoor setting, without changing film or camera bodies. What's more, some of the modern image sensors are functional at extremely high ISO numbers, like 64,000 or more. This lets you capture unique shots in very low light. Additionally, since the image sensor on some cameras is used for both photographs and for videos, altering the ISO of your camera affords you video in challenging light as well (Image 10.2).

Textbook Lighting for Great Footage

Fairly bright, flat light is the best environmental condition for taking great video shots. Why seek flat light? Let's start the conversation by talking about the qualities of light. Contrast in film refers to the difference in light intensity between bright and dark areas. Shots composed of very light and dark regions have high contrast. High contrast shots make it difficult for viewers to see details in the bright or dark regions. Details are lost because cameras on automatic light adjustment settings try to balance the lighting for both regions. This typically results in the bright areas being overexposed, blown out, and colorless, while the dark areas are cloaked in darkness. Both of these make for poor quality shots (Image 10.3).

Great flat lighting happens on overcast days. On such days, you can tell where the sun is in the sky by it's a faint glow behind the clouds. At the same time, the clouds serve to diffract direct light and reduce harsh shadows. A very simple and reliable test for good, flat light during the day is to be aware of your facial expressions. Are you squinting, covering your eyes, or straining to make out details as you look at your subject? If no, you stand a great chance of reliably filming quality shots.

Image 10.2 Turning up the ISO on your camera increases a camera's sensitivity to light, yet reduces the camera's ability to reproduce colors. For our first shot the ISO was set at 1600. The shot was dark yet the colors are appealing. We then increased the ISO to 6400 for the second shot. This brightened the scene, yet was accompanied by increases in granularity

a

b

Image 10.3 (**a**) Cameras are not as good at accounting for high-contrast lighting as our eyes. However, in automatic light setting modes, they do their best to compensate for bright and dark areas within a scene. Sometimes failing at capturing shots with great potential. (**b**) Flat light usually makes for better composition. Deep shadows or areas with excessive light give way to more details, at the cost of decreasing the apparent dimensionality within a scene

If you are not in flat light, seek it out. What could this look like? For a simple example, wait for the sun to go behind clouds when shaded spaces are not easy to find. Stop filming when the sun reemerges. If filming between periods of intense sun becomes frustrating, seek out more permanent shade. How about on downtown streets of a city? Avoid direct light by finding areas where the light from the sky bounces down from glassy walls to the streets below. What if you are conducting an interview? Lead your interviewee out of the direct sunlight to behind a building or into the shade of a richly vegetated forest.

The same approaches stand for indoor shots. Watch for individual lights beaming directly down from above. These cast harsh shadows. Find a room where there are several lights, as opposed to one.

Personalizing Your Camera Light Settings: Automatic Settings

The simplest way to brighten or darken a shot, as seen through your camera (without using external lighting equipment), is to set your camera to *automatic* mode. Automatic settings are usually indicated by a green *A* or *Auto* on your camera body (Image 10.4). Cameras are becoming smarter and smarter at evaluating what is happening in a scene and making internal adjustments to draw out as much detail and balanced-color as possible. Automatic settings instantaneously adjust shutter speed, f-stop, and ISO to make this happen.

As mentioned before, shooting on automatic has its pitfalls, and a few variables can confuse your camera away from the outcomes that you would like. For example, a wee bit of light in the corner of your shot might force your shot to darken (Image 10.5). If the opportunity presents, turn the camera away from the source of light. This slightly changes your framing, yet improves the lighting. With some cameras, you can turn the camera away from that light, lock in the light settings directed at a space characterized by better lighting, and then turn the camera back to the original framing and record. You will have to read your camera's manual to understand how or if you can lock in the light settings.

Cool Trick
Some cameras allow you to tell it, in advance, what type of shot you are hoping to capture. Landscapes, sports, portraits, etc. If you designate that you are photographing sports some cameras take measures to increase the shutter speed so that images have reduced blur associated with fast movement. Alternatively, it will change the depth of field (blurring out the background) when you shoot portraits. Incredible!

Image 10.4 Some cameras leave you very few options for manipulating how you control light within your shots. Others give you more. Here we labeled the control wheel on a DSLR. By turning the wheel you access different tools for controling lighting, depth of field and, to a smaller degree, crispness of shots

Personalizing Your Camera Light Settings: Priority Settings

While your camera's electronics are quite sophisticated, you can make changes to the settings (like shutter speed) on your camera to enhance the quality of your shots. This means that you become more active judge of the lighting in your shots. What's more, you will have more power to improve shot characteristics.

With an understanding of aperture, shutter speed, and ISO, you are ready to take greater command of the light settings on your camera. Here is a refresher on why. (1) Camera automatic settings get confused in high-contrast settings. (2) Precise light adjustments are fantastic tools for highlighting the most important pieces of information in a scene. (3) Lighting is a photo/videographer's paintbrush and creates provocative shots. Let's talk about how you can change settings to manifest the effects that you seek.

A good place to start, when taking over the controls of your camera, is with modes that give governing priority to either shutter speed or aperture. In priority settings, the camera will still attempt to optimize your shot for lighting while either your aperture or your shutter speed is locked. When you choose to lock in your aperture, the mode is called *aperture priority*. *Shutter priority* locks the shutter speed. The camera will then correct the settings that are not locked to make the shot appropriately bright. This way you can shoot action with high shutter speeds or open your depth of field by shutting down your aperture without thinking too hard

Image 10.5 The automatic settings on many basic cameras struggle at making appealing light adjustments when pointed at bright lights. Oftentimes, they darken everything else within the scene to compensate for the one bright light source. Some cameras retain light meter settings directed at dark areas, thus brightening a shot and drawing out features in shadowy regions

about having to evaluate the light conditions. Most DSLR cameras and higher-end camcorders identify these modes with an "A" or "S" on the control wheel of your camera. Refer to our earlier image of a camera's control wheel.

Exposure Compensation

Let's go a little further into what you can do with aperture and shutter priority settings.

In the shutter priority setting, you lock in the amount of time that each frame is exposed to light. Maybe, 1/60th of a second. Now your camera will automatically adjust the aperture to make for what it thinks is great lighting. What if the shot appears too dark and you want to brighten your scene? An option called *exposure compensation* will permit you some freedom to do just that. On many cameras, the exposure compensation setting looks like "+/−." Selecting this option while in shutter priority, lets you shift your camera's automatic aperture settings towards those that will improve your shot. This is a great mechanism for overriding the automatic settings to slightly brighten or darken your footage per yours.

Similar adjustments are also possible in aperture priority mode. In these situations you have locked in the size of the aperture. So shutter speed will adjust to make for what it thinks is a well-lit clip. If you then make exposure compensation changes, brightening or darkening will be driven by changes to shutter speed. The more time, the more light, the brighter your shot. The less time, the less light, the darker your shot.

Cool Trick #1
While you can adjust your priority settings to aperture or shutter speed, ISO is still a very viable option for changing your camera's sensitivity to light. If you want a slower shutter speed, but the aperture settings on your camera won't close down any further and keep your shot to the right levels, consider manually changing the ISO settings to get you there. Again, higher ISOs brighten your scenes and lower ISOs darken scenes.

Cool Trick #2
A tidbit of advice as you take manual control of light settings on your camera: adjust so that the most important elements are in ideal light. Find the elements that will carry your story most forcefully, say a person's face, a dial on an analyzer, or trees swaying in the wind, and show these in exquisite detail.

Warning

As a warning, some cameras will hold your previous exposure compensation settings until you change them back. You might have made perfect adjustments for one scene, yet these settings might persist until you take action to change them. Be aware of this issue and be vigilant to reset exposure compensation after each shot.

Personalizing Your Camera Light Settings: Fully Manual

Taking full control of your camera is often designated as "Manual" or "M" on your camera's mode toggle or wheel. In this mode, you can precisely dial in the shot that you want by adjusting the shutter speed, aperture, and ISO at the same time. On a first order, having control over all adjustments allows you to brighten or darken your shots over an enormous light intensity spectrum. This makes filming in light that challenges automatic settings much easier to manage. Manual mode literally puts all adjustments on your shoulders. Perhaps manual sounds too complicated. Are you are on the fence about how far you can go with manual settings. A day of tinkering with aperture, shutter speed, and ISO, simultaneously, can make you dangerously good at working with variable light conditions and creating different tones to your scenes. Be warned, these experiences have sucked a large number of people into a lifetime devoted to the art.

> **Cool Trick**
> As we said earlier, in areas where your lighting has very high contrast, balancing light can be very challenging. It is difficult, if not impossible, to show details within the shadows as well as in the light (larger image sensors are better at this than smaller). One solution to high-contrast light is to use its harsh dynamics to your advantage. First, high-contrast lighting is an effective tool for drawing out very important elements and forcing others to fall away. Consider turning your most important element into a silhouette, through exposure compensation or manual adjustments. Silhouettes are striking shots, because they ask the viewer to see familiar objects in new ways. In these cases, light metering means that you are vigilant to make your darks very dark and your lights well balanced (Image 10.6).

Changing Native Light

We just discussed a number of very powerful ways to control the settings on your camera or the framing of shots to make for better lighting. Manipulating the light native to a setting will also improve your shots.

When lights in a room are too direct, perhaps resulting in deep and distracting shadows, some videographers add diffusers. A diffuser takes direct light and flattens it through scatter. Diffusers can be as simple as a thin bed sheet placed in front of a light. Other diffusers include sheets of opaque plastic. Be aware, the act of diffusing light may increase the quality of light yet in the process decrease the amount of light reaching your subject matter.

If you are in light-limited areas, working with native light and diffusers can present a crux. Consider importing light. On the go, I bring a couple $30 mobile shop lights and extension cords. Modern lights are very bright, quite small and do not burn too hot. As a result, a room packed with cameras, analyzers and people do not become unmanageably hot. As another benefit, the cooler burning means that you can place a simple diffuser practically on top of the light. Another solution to reducing the directness of their blaring light is to aim the shop lights away from the subject matter and towards a nearby white wall (if they are present). The paint serves to refract the light, making for flat light all over the room. The difference between this application and direct light is incredible! (Image 10.7).

Walls are not the only way to reflect/refract light. For as little as $20, foldable *bounces* are available for purchase at stores dedicated to photography and video. These reflectors serve the same diffractive purpose as a white wall; however, they can also double as a reflector, steering light from one area of a room to spaces where it is dark. For example, if you are conducting an interview with numerous ceiling lights casting deep shadows under a subject's eyes, use the reflective dish to redirect some of this light more horizontally, thereby filling out the dark shadows. Unfortunately, these reflectors are not very easy to orient without a stand or an assistant. If you have a friend or a colleague that wants to join in on a shoot, put them in charge of a reflector.

Image 10.6 When filming in high-contrast settings, work with what you have. In this case, there was no way to retain the details of the mountain in the background and capture the colors of the climber in the foreground. As such, I chose to keep the background lighting well balanced, while turning the foreground into an appealing silhouette

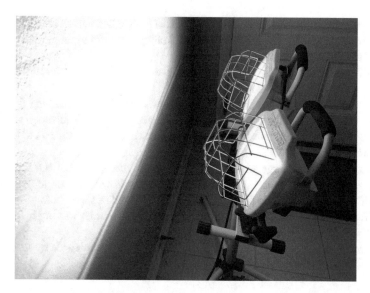

Image 10.7 Point-source lamps create very harsh and contrast-rich lighting. Directing such lights onto subject matter is often overwhelming. Consider redirecting the lamp towards a wall, such that the light refracts into numerous directions, thus softening the light conditions

Are you someone who will dedicate space to projects such as interviews? Professional-grade photo and video lighting is worth the investment. They have been designed with your needs in mind, and their adjustability makes them incredibly user-friendly. Quickly, you can control the directness and direction of the light. As a consequence, you are adding more illumination power, increasing your? flexibility to close the aperture on your camera(s). Again, the associated increases in depth of field may save you time and effort when filming very active subjects. Let's say that an interviewee bobs and weaves – a closed aperture setting will keep them in focus for more time than with limited light. Note: many professionals swear by certain depths of focus for interviews. More lighting may accommodate these choices as well.

Breaking Down What We Learned About Lighting

Now the power is in your hands to manipulate the shutter speed, aperture size, and ISO to optimize the clarity, lighting, color balance, focal range, and composition. Unless you have previous experience with cameras, all of this may be a little overwhelming. Here are some simple reminders of what we just discussed.

- One of the easiest ways to manipulate the light on subject matter is to move the objects of interest into better lighting.

- Unless you are specifically looking for areas of high light contrast, seek out flat light.
- Fast shutter speeds are great for photographing rapidly moving objects without showing a consequential blur.
- Slow shutter speeds allow more light in, thereby facilitating the closing of your aperture and increase your depth of field.
- A wide aperture (small f-stop number) lets in more light, brightening dark shots. This also gives you the option to increase shutter speed. Unfortunately, wide apertures reduce your depth of field, meaning you have to be more vigilant to keep your shots in focus.
- A small aperture (large f-stop number) reduces the amount of light hitting the image sensor. This increases your depth of field, allowing a simple focal adjustment to resolve both the foreground and background of your shot.
- A low ISO setting means that you need more light in order to film; however, the quality of your clips improves.
- High ISO settings permit easier filming in low light or high activity settings, yet manifest in more granular clips.

Chapter 11
Making Visual Composition Dynamic

Focus

Effective videography hinges on taking extra care that the essential elements in your scene are in focus. Clips that are slightly out of focus can't be recovered once in the editing studio, and immediately stand out as careless faults in your film (Image 11.1). In most cases, audiences have a negative reaction when they spot poorly focused scenes and immediately identify the work as unprofessional or not important enough for their full attention.

Lots of videographers depend upon the autofocusing options on board just about every video camera. Each camera focuses on different grids or matrices in order to sharpen your shot. Some may pick a number of points scattered throughout your scene. Other cameras focus on faces or pick the center of the shot. The matrices that cameras use for focusing can be manipulated to best suit filming applications. Some people are exceptionally proficient at working within these settings for great success. If you want to depend upon automatic focus options, explore how your camera operates in these differing focal modes.

Unfortunately, automatic focus technologies have their weaknesses. In these instances, you can lean on your greater knowledge of how cameras work to focus on what you want. For example, in low light your camera commands the aperture to open wide (to allow in the most light). In review, as the aperture opens (lower f-stop numbers) cameras struggle with keeping what is in the foreground and background in focus at the same time. As such, in dark environments a camera set on autofocus mode may not know what to focus upon – objects in the foreground or background? Focus will jump back and forth resulting in poor-quality video clips. This problem becomes more pronounced as you pan your camera over a scene with objects at different distances from your camera.

Image 11.1 Crisp focus is a must. Shots that are slightly out of focus are quick to frustrate audiences

The first solution to this conundrum revolves around methods to expand your camera's depth of focus. How might you do this? One answer is to brightening the scene by either adding supplemental lighting or moving the subject into a brighter space. Increased light affords you the chance to close your aperture and expand your depth of field.

Other options include the manipulation of the settings on your camera. First, slow your shutter speed down to 1/30th of a second (one of the standard frame rates at which cameras record video). This allows the maximum amount of light to illuminate each frame before you start creating seriously deleterious effects to your footage. Then decrease the size of your aperture (higher f-stop). Again, this reduces the light hitting the image sensor, but expands your focal range.

Additionally, consider augmenting your image sensor's sensitivity to light, by increasing the ISO. In short – the higher the ISO, the less light you need to capture a scene, and the more rapidly your camera can record color and light. As a consequence of higher ISO, there are attendant losses to the color vibrance of your shot. Despite gradually decreasing color qualities, higher ISOs may give you the wiggle room to close down your aperture to a smaller size.

If these steps do not give you freedom to increase your depth of field, you will have to take action to optimize shots that are limited by a shallow depth of field. I start by directing my camera at the elements that I need to stay in focus, and allow autofocus to bring resolution to those elements. If my camera comes with a focal lock, I activate it, or simply turn my autofocus off. That way, if I move the gaze of

Image 11.2 In order to decrease your depth of field, you must open the aperture on your lens. Doing so draws viewer's attention to important details

my camera away from that element, the focal setting will be maintained. Less-expensive camcorders might not have the option to lock or turn off auto focus. In this case, be very careful to not have great differences of depth of focus between moving elements. This is particularly true in dark scenes.

Oppositely, shallow depths of field make for some very cool composition. In fact, learning how to work with short focal ranges is a revered art that takes a trained eye and camera operator. For example, shallow depths of field steer attention or give impact to specific elements in a scene. I often turn to this technique when filming interviews, portraits, or very specific gadgetry. In the photo above, a grasshopper on a screen pops out, while the background complexity blurs into a mist. If the depth of field were larger, the grasshopper would have been lost in the background (Image 11.2).

Cool Trick #1

A shallow depth of field within a clip does not mean that you have to maintain focus on foreground or background elements throughout a shot. With a steady hand, you can change the focal length from close-up to far away (or vise versa) as you record a scene. This draws the viewer's attention from one element in the scene to another, thus enhancing how you tell your story. Think of this – the opening to a scene begins with a person in the foreground staring over a wall. The background is out of focus. As the clip progresses, the region of crisp focus changes from the person to a lion in the background. The story of a person watching the lion unfolds in a richer way. This technique is called "rack focus" (Image 11.3).

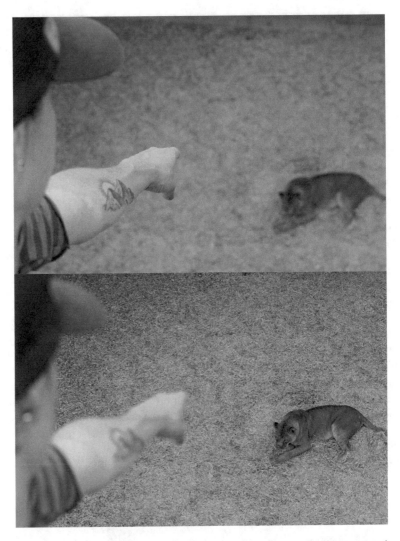

Image 11.3 Changing the focal range of your camera throughout a clip transports a viewer's attention from one element to another. In this case, the audience's attention is drawn from the person watching a lion, to the lion

Cool Trick #2

Let's talk about how short focal ranges enhance slider shots. I find the impacts of a slider (discussed in Chap. 6) particularly riveting when a videographer sets the aperture size to a very low f-stop. With the foreground (background) is in focus while the background (foreground) is not, push the camera through a setting - objects pass in and out of focus! This is a fantastic way to draw the eye towards important themes in a shot sequentially.

Zooming

Zooming uses lenses to transport a viewer closer to elements in a scene rather than the videographer picking up the camera and moving it closer. Zooming is a quick way to change the framing of your scene between shots. If you have the right lens, you can change the framing from wide, mid, and close-up in very little time.

Practically every camcorder gives you a couple of speeds for zooming in or out. It is rare that the rapid zoom setting makes for decent, useable footage, unless this zoom is the precise feel for which you are looking. Slow zooms are typically more useful, making for easy transitions from or into greater detail.

As with most camera adjustments, zooming comes with a few complexities. Firstly, as most cameras zoom in (zoom out), the barrel of the lens extends (retracts). This act changes the amount of light that hits the image sensor. If it were not for camcorders making real-time adjustments to their aperture and shutter speeds, zooming in (out) would make a shot darker and darker (lighter and lighter). Since aperture settings change incrementally, a careful eye can catch these step changes during a recorded zoom. For some, these adjustments may not be appreciable, yet others are much more obvious. The first step is to be aware of the potential impacts and watch for where they may detract from a shot.

Many DSLRs struggle to compensate for the changing amount of light that hits the image sensor as you zoom in or out. Why? Some DSLRs, even set in automatic light-adjustment mode, do not make real-time corrections for light while recording a scene (light settings often get locked in before you start recording a scene). This does not mean that all is lost if you hope to record during a zoom with a DSLR. A simple work-around for this condition can be performed back in your office with your video editing program. Higher-end editing packages allow you to brighten (darken) a scene defined by zooming in or out.

Complexities compound. High-end, stock DSLR camera lenses rarely come with a motor drive for smooth zooming. It is very difficult to manage a smooth wrist movement to produce a consistent zoom speed. Inevitably your hands will change where you put force on your lens during a zoom, manifesting as a slight bobble in your scene. Solutions include purchasing a motor drive (not a small expense) or practicing your smooth zooms until the bobbles are imperceptible.

Panning Horizontally

Panning is the act of swiveling a camera to capture a greater scope of what surrounds you. Most often this includes turning the camera horizontally on a tripod head to trace along a horizon. However, one can also pan up and down.

Horizontal panning is fantastic for keeping horizontally moving objects centered within your scene. Perhaps you are following a runner in a marathon. Maybe you

Image 11.4 Should you be presented with a wide scene, a wide-angle shot will capture every-thing; however, you might lose the details that you want. Instead, place your camera on a fluid head tripod, or monopod, zoom in a little closer with your lens, press record and swivel your camera to capture the complete scene. This act is called *panning*

are capturing falling dominoes (from the side). Alternatively, horizontal panning is a good method for capturing the details of an object that is wide and not very tall. I am thinking of a mountain range. A wide-angle shot would mean that you see the breadth of the range, but not the details of the mountains. Oppositely, should you zoom in, you only capture a small subset of the mountains. Panning through a landscape will record the details and the breadth of the mountains in the same clip (Image 11.4).

Just like zooming, the best pans are slow and steady. While camcorders come equipped with motors for gentle zooms, panning motor drives are not standard for cameras or tripods within the purchasing brackets that we are discussing. As such, learning how to couple movement of your body with those of your tripods is paramount.

I would say that a first step is to practice panning more slowly than you might think is necessary. For many, the natural inclination is to pivot too fast. Slow pans give viewers the chance to soak up the details. Additionally, just like for slow motion clips, a slow pan can lend a feel of tension, drama, and importance. As an experiment, try panning at what you think is a good speed, then put in another attempt at panning that pivots that camera at half or a third the speed that you think is adequate. See what you think. Even if pivoting is controlled by following a moving object in your field of view, step a little further back from the subject matter so you don't have to pivot quite as fast.

Liquid Head Tripods

Erratically paced pans are a sure way to give a movie an unprofessional feel. Additionally, rapid movements with cameras that have sensitive rolling shutters (such as DSLRs) create undesirable secondary visual stuttering. The best tool for making slow and smooth pans is a fluid head tripod. We mentioned some initial thoughts about fluid head tripods in the chapter dedicated to buying stabilizing production gear. In review, fluid head tripods do not have fluid in their head as much as they feel very fluid as you pivot. They produce just the right amount of resistance so that pivoting is not disrupted by small variations in speed. The result? A liquid-smooth pan.

I try to keep four thoughts in mind when pivoting my camera on a fluid head tripod.

- First, I stabilize my feet. A very secure body position means that I can focus all of my energy on using my arms, waist, and shoulders for a very slow and even movement.
- Second, I don't watch my scene with my face close to my camera's viewfinder. Instead, I situate my head in a place further back where I can see the viewfinder at the beginning and end or my pivot (if possible).
- Third, pivot the head with even force. Some fluid heads rotate effortlessly; however this is not always the case. When there might be hints of tightness, I use as much of my body mass as possible to force a smooth rotation.
- Fourth, try to pan with a wide-angle lens setting. Zooming in exacerbates any bobbles or erratic movements in my pan. What's more, landscapes pass too rapidly through a scene under zoomed-in settings.

Non-fluid Head Tripods

Non-fluid head tripods do not make for consistently paced turns. For example, with a very inexpensive tripod, one can feel the tension periodically increase in a pivot, followed by quick releases.

If you're stuck with a non-fluid head tripod, and still want to make smooth panning shots, what can you do? First, grab a monopod or convert your tripod to a

Image 11.5 One can
deliver on effective pans
without a fluid head tripod.
As opposed to pivoting a
camera about the tripod
head, collapse a tripod
down to one leg (or use a
monopod) and pivot the
camera about the singular
point connecting with the
ground. Use your body
weight to keep the pivot
smooth and regulated

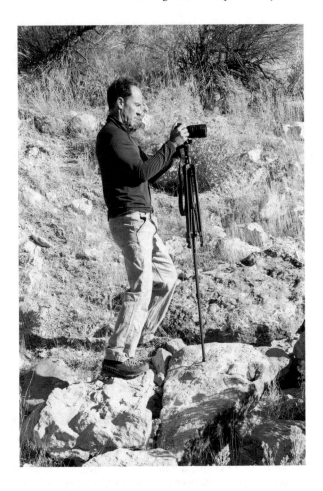

monopod. You accomplish this conversion by collapsing two of the legs and then propping your camera on top of the only remaining leg. Level out your shot with the single leg standing as vertically as possible. You will be using the hard iron point at the end of the single leg as your point of rotation upon the ground (instead of the tripod head). Your job is to maintain the camera's upright positioning as you turn the entire assembly (camera and tripod) with your body. I would suggest cradling the camera with both hands while tucking your hands up close to your body, thereby stabilizing its position with your elbow locked against your rib cage. Then rotate your body and the camera about your hips while keeping your feet firmly in place (Image 11.5).

Another option for panning with a monopod is to place the pointed end of the monopod between your feet (shoulder width apart). Then lean the camera out to a full arm's length away from your body. Now level up your camera so that it will capture the horizon that you wish. Lock your arms out straight and pivot about your

Image 11.6 To build upon the first method for panning with a monopod, you can swing a camera about your body's center of gravity much like a pendulum. Indeed, this pan pivots the camera at a larger radius than that described in the previous image, and thus interacts with the field of view differently. Try both methods and explore how your camera passes over the various elements within your pans

hips and shoulders, swinging the camera like a pendulum. Because the weight of your camera is at a wide this techniques makes for effective pans. However, it is a bit harder to control the speed and levelness of rotation (Image 11.6).

Without a Tripod

Pans become much more difficult when you are without a tripod. If you find yourself in this situation, think about how you can use your body as the center of rotation and the tool for maintaining the camera's height. What's more, it must create a smooth and stable pivot. Quite the challenge! I would say that the following solution is not universally accepted, but it has worked for me. Take great care that your feet are firmly planted on the ground – about a shoulder width apart.

Image 11.7 In a pinch, one can produce acceptable pans without a tripod. For this, you want to create a stable arrangement between yourself and the camera. Lock in your arms and consider using a camera strap to hold tension against your body. Emphasize pivoting about your hips

Lock your legs such that your hips are level to the horizon. Once again, your torso will serve as the swivel point for the pan. If you are dealing with a camera that is typically held in front of your face, like small camcorders, bring your hands in close to your chest, thereby firming up your arms and use your core to smoothen your pivots. I put on the neck strap over my head and lightly push the camera away from my body with two hands (elbows locked on my rib cage). This ads tension in the strap and steadiness to the camera, while giving you space to watch the viewing screen (Image 11.7).

Image 11.8 One can exact vertical pans without a fluid head tripod. With the use of two tripod legs in contact with the ground, pull or push the camera towards or away from your body. Long tripod legs make for smoother, yet shorter-angle pans, while collapsed tripod legs make for less-smooth, yet larger-angle vertical pans

Panning Vertically

Panning could also mean that you wish to trace a scene from up to down or from down to up. Exacting these shots is easily achieved with a capable fluid head tripod. I find that as long as you stabilize your body, and weight the tripod such that the legs won't lift as you tilt your camera, vertical pans are as easy as horizontal pans.

If you work with a standard head tripod (not fluid head), don't expect the tripod head to help you. However, you can still achieve a smooth vertical pan with some ingenuity. Begin by changing your non-fluid head tripod to a dual pod by collapsing one leg down completely. Now set your two remaining tripod legs apart on the ground, aligning with the axis that you would like to hinge your pan along. This means that the points for the two legs will serve as a fulcrum for the pan. Step behind the camera and stabilize your legs (or kneel), such that your core is steady and you can see the viewing screen of your camera. Tighten the tripod head settings into the orientation that captures your starting scene. Now simply extend or retract your arms, panning the camera down or up efficiently. The longer the tripod legs, the smoother the pan; however, longer legs will decrease the pan angle that you can achieve between fully extended and retracted arms (Image 11.8).

As we asked with horizontal pans, is it possible to manage a vertical pan without a tripod? I find these shots more difficult to capture, and the measures that I take for executing these vertical pans awkward. As always, begin by stabilizing your legs and core. I would suggest positioning your feet so that they are shoulder width apart, with one beside the other. Even better, kneel, because it is more stable. This exercise

Image 11.9 Vertical pans without a tripod are very difficult to execute. When necessary, hold your camera close to your chest, stabilize your legs, and pivot up or down about your hips. This exercise can be done in a standing position, but is best conducted while kneeling for more stability

is best done using your hips as the hinge point for pivoting up and down. If precise framing of your shot is not critical, you may wish to hold your camera to your chest, with your elbows locked at your ribs. If seeing the viewfinder is a must, wear the camera neck strap and continue to pinch your arms against your ribs, but hold the camera as closely to your face as you can without losing eye contact with the viewfinder. This closeness is useful in that the further that your arms are away from your body, the more hand-waving will affect your vertical pan. Press record and pivot at your hips. In order to get a little more rotation to your pan than your hips can afford, hold the camera closer to your face, and couple bending at your hips with craning your neck. This can afford you 15° more of pivot. Again, since your elbows are pinned, the weight of your head and the solidified hinging point can still make for decent shots. Articulating your wrists to add more rotation is a crapshoot and not suggested (Image 11.9). If your shot pivots steeply downward, be aware that your neck strap may enter into the shot.

Manipulating Time

Speeding up or slowing down film produce some very evocative media. For example, when a half-hour clip of a thundercloud is compressed to just a few seconds, there is immense tension built around how extreme the weather is. Oppositely, the classic slow motion shot of a boxer being punched in the face steeps the viewer in prolonged movement of sweat, saliva, and contorting facial expressions. Preparing these kinds of shots takes some forethought and careful completion.

Time-Lapse Video

Time-lapses are scenes that are sped up, maybe two, ten, or a hundred times faster than reality. Compressing a scene, say from an hour of footage to just a few seconds, changes the way that viewers perceive realities. Scenes that would otherwise seem tragically stagnant are brought to life.

The first method for filming a time-lapse is quite easy, yet uses a large amount of camera and computer memory. Begin by setting up your camera on a solidly positioned tripod. Seek out areas that are not in the wind as it can send vibrations through the tripod making for shaky time-lapses. Additionally, a stiff gust could shift, even topple, a camera.

Cool Trick
When preparing to film a time-lapse shot, tighten all adjustment knobs so that tripod legs don't sag or the tripod head does not droop over time. Inexpensive tripods are notorious for slumping when a heavy camera is placed on top (ruining what you thought was a level shot). I find that these issues crop up most notably with small and plastic adjuster knobs used for leveling a camera. Attempting to execute leveling a heavy camera on top of a flimsy tripod head is often futile. Instead, loosen the leveling knob and allow the camera to settle into whatever position it is most stable – which may be 10° off of level. Now tighten the knobs, locking the camera into its unlevel position. Lastly, level the camera by changing heights of the legs. Yes, this might destabilize the tripod some; however, if you only need to adjust your camera by 10°, steadiness should be maintained. This technique is a simple, yet hardly elegant solution for making for micro-leveling adjustments.

The next step is to lock in focal settings on your camera. Attention to this step is less important in very controlled environments such as a laboratory, but very important when you film outdoors. Why? Little things can draw the focus away from your core elements. For example, let's say that you are filming people milling about in a park, yet a fly or raindrop lands on your lens. Autofocus might trend away from the park and towards the fly. Your shot is ruined. Clouds in low light are notorious for confusing the autofocus on cameras. Cameras end up searching for any object that is in focus, thereby blurring entire scenes. It is awful to review your clips once back in your office and find that focus was lost several times over the course of a time-lapse. As such, if your camera presents the option to lock in focus settings, lock them in on your desired elements.

If you want to keep a certain element bathed in consistent light, say a new oil-drilling rig positioned on a hillside, it is useful to turn to automatic light adjustments on your camera. That way you can maintain a viewer's attention on the rig as the setting might change around it. However, this method is not always effective when you wish to emphasize a setting rather than an object. In my experience, this is, more often than not, the reason for shooting a time-lapse. Instead, lock in the light settings on the camera. Let's talk about why from the hypothetical scenario of you wanting to speed up a shot of swirling clouds. It is often the case that swirling clouds will darken and brighten a setting several times over 10–20 min. If this scene were filmed on automatic light settings, the camera would compensate for dark periods by letting in more light. Likewise, the camera will let in less light when a scene is bathed in brightness. Because your camera is reacting to light levels, the brightness in your resulting clip will remain somewhat constant. Usually this means that dynamism lost. Oppositely, consider when the same scene is filmed with locked light settings. During periods of light, the scene will explode and during periods of low light, it will dive into darkness. The latter encapsulates the wild energy embodied in the clouds.

Once you have framed, focused, and adjusted for the light in your shot, press record.

Filming time-lapse shots occupies your camera for an extended period of filming time. This begs a very good question. How long do you let your camera run in order to get a good shot? Unfortunately there is no easy answer. It greatly depends upon the process that you are filming and how long you want the process to appear when embedded as a clip on your final video. Let's say that you want to film some form of sample analysis – maybe a robotic arm picking up a sample, moving it to a new location, and having a new sample follow into place. At the very least, I would capture the entire process plus some extra time on either side of the process. This means that you start the filming well before the cycle begins and let it run well after it ends. Doing so leaves you with additional footage to introduce or conclude the scene.

Everyone has his or her own taste for the speed of time-lapse videography. As a general rule of thumb, I do not like to film clips that are sped up by a factor of around 2 or less. Such clips seem forced and awkward – so close to real time, but obviously not. Additionally, I do not like it when too many important details are shoved into one time-lapse. If more than a few elements are moving at speeds that the human eye is not used to, time-lapses can be overwhelming. Indeed, this could be your mission, so go ahead and film as many moving parts as you feel will serve your purposes.

Note 1 The more wide-angled your shot, the slower the movement will appear in a time-lapse. Because of this condition, I generally film with wide-angle settings longer than if it were to be zoomed in.

Note 2 Many cameras cannot film in high definition for more than 12 min without creating multiple movie files. Longer time-lapses will require you to stitch files together later on in your editing studio.

Time-Lapse with Photos

What if you plan on filming a time-lapse over long periods of time (in excess of 30 min)? It can be useful to use a camera that is programmable to take photographs at regular intervals. Such cameras compile a collection of sequential photos that, when linked together with software, produce breathtaking speed-up scenes. While not all cameras come stock with this option, more cameras facilitate interval shots with third-party devices. Even more exciting, one might be able to download apps to serve this purpose. Read up to see if your camera affords you this option.

For those who are lucky enough to have a camera that shoots at regular intervals, you can turn a pile of sequential images into a time-lapse. This begins with figuring out how long you want your time-lapse to be, in conjunction with knowing the frame rate that you wish for your resulting film. Let's say that your video will play at 30 fps and you would like your time-lapse to last 10 s. Do the math – you will need to stitch together 300 frames to create the desired time-lapse. Now, how long do you think you will film? Maybe two and a half hours, which also amounts to 9000 s. In order to film 300 frames in 9000 s, your camera must take an image every 30 s. This is your target interval!

Cool Trick
There are benefits to building a photo-based time-lapse from high-resolution images. High-resolution photography means that you can crop in to smaller zones, and the images suffer no loss in the resulting video resolution. Now you can zoom and pan across a time-lapse as the scene unfolds. As an example, let's say that your photos capture a radar dish station situated on a hillside. After you have filmed your sequence of images, you could start your time-lapse video clip zoomed in on the station and, through time, zoom out to show how the facility is nestled deep in a tropical forest, surrounded by wild, swirling clouds. This type of edit is an advanced step that does not come standard with all photo/video editing platforms, but this example does generate a great conversation about the exciting outcomes that you can deliver from shooting time-lapses with photographs.

The photos that most cameras capture are of much higher resolution than those of the video that you hope to create. For example, the photo option of my favorite camera is 42 megapixels. Huge pictures! If I use these high-resolution photos while I capture an interval-based time-lapse, I will use up an enormous amount of space on my memory cards and hard drive. What's more, the assembly and export of such files will bog down my computer processor. To work around this potential bottle-neck, set the photo resolution of your camera to slightly greater than the resolution of the desired video. For example, I may want my final video to be 1080p, which means that the clips have 1080 pixels in the vertical direction and 1920 in the horizontal. So, I set my camera to shoot photos that are closer to 2000–3000 pixels across.

The act of stitching your list of photos together used to be quite entailed. Fortunately, modern video and photo editing programs provide users with options for assembling your clips rather effortlessly. The key to working with these programs is to make certain that your camera names images numerically, in order. With this requirement in hand, you can rapidly assemble great photo-based time-lapses. This is truly a magical method that modern technology affords filmmakers.

Final Thoughts on Time-Lapses

Regardless of how you approach time-lapse videography, the exercise drains battery resources. In fact, some cameras struggle to last long enough to finish desired time-lapses. If you cannot purchase a battery that will take you from beginning to end of a time-lapse, it is helpful to determine whether you can plug your camera into a wall outlet and thus gain long-term power.

Slow Motion

The opposite of a time-lapse is slow motion video. This is when you slow down your footage such that it lasts longer in your film than the original clip. Slow motion shots are great for steeping a viewer in the action of a scene. Not only do they have more time to process details, but it also changes the way that they interact with or understand an action. A ski jumper seems to soar effortlessly through the air, and the audience is given a sense of flying. In science, slow motion opens windows for perceiving action that is otherwise lost in real time. Take a look at slow motion videos posted on YouTube showing a balloon popping, a rock falling into water, or a golf ball being hit. Balloons appear to shatter, water molecules glom to each other, and golf balls compress to a fraction of their original diameter! All of a sudden, the physics at the foundation of these two processes become more apparent and hopefully easier to understand (Image 11.10).

The default frame rates of most cameras are set at standards, 30 or 24 fps. These frame rates play very smoothly at their intended rates. However, if these scenes were slowed down to half speed, the smoothness is compromised, giving clips a

Image 11.10 Slow motion video is a powerful tool for changing the way that people perceive normal processes. It slows action down such that new details unfold to the viewer. Here a rain droplet fights surface tension of a standing body of water

strobing/stilted feel. In some cases, this can make slow motion scenes more distracting than useful.

So how are smooth slow motion shots made? Standard frame rates are not typically set for filming with the intent to make slow motion clips. However, many higher-end cameras give you the option to switch the frame rate to double or quadruple the normal 30 fps. If you film at 120 fps, you can slow your footage down by 4 and still retain a film speed of 30 fps in your final movie.

Why are cameras not set at higher frame rates, thus facilitating a larger number of slow motion shots? First, as we discussed earlier, the speed at which scenes are filmed aligns with standard rates used for television, cinema, or Internet broadcasting (either 24 or 30 fps). Second, faster speeds require rapid camera processing and more memory space, both of which encumber filming and editing. For some computers, this might mean that you won't be able to watch your edits without your computer hanging up from time to time. When you are editing with time efficiency in mind, these hitches can be incredibly aggravating.

As mentioned earlier, speeding up the frame rate of filming stresses the processor of cameras. To compensate for that processing, many cameras decrease the resolution at which the faster footage is recorded. A camera that normally films 30 fps at 1080p, might film 120 fps at a resolution of 720p or less. Be aware that your camera might do this, and resulting footage will not have the quality of shots filmed at a standard rate.

Note The technology and tools for filming slow motion clips are rapidly growing. As such, slow motion options abound for little to no cost. Check to see if your smartphone has slow motion capabilities. What's more, these options come stock with most modern sport cameras (like GoPro).

Cool Trick

What if you really want or need to slow down a shot that was filmed at 30 fps? By slowing the clip to, say, 10 fps, you will likely observe stilted footage. Third-party editing programs can help you out. For example, companies have developed ways to interpolate between each frame to smoothen out the roughness. The scenes are still missing critical information that you can only capture when shots are truly filmed at high speeds, but the outputs from these programs are impressive and might meet your needs without you having to purchase another camera.

Final Thoughts on Filming

Let Shots Linger

As you film, record your scenes a little longer than what you think is adequate. Let's say that you are taking a heap of B-roll within a laboratory. It is generally accepted that the capturing of a B-roll shot should occupy about 10 s of time. This gives the camera time to stabilize (if held by an inexpensive tripod) and opens the option for the editor to show the scene and add in a gentle fade to the next scene as you are piecing together your story. Particularly with filming people and wildlife, I think back to how many times I have wished that I let the camera roll a little longer. In my haste to capture the next shot, I missed something really catchy. The one that comes to mind was a perched eagle. A blink after I ended recording it opened its giant wings and took flight. Amazing! Two more seconds of recording would have delivered on a shot of a lifetime. Unfortunately, time passes very slowly as the camera is running – three seconds can seem like 10. As a remedy to slowly passing time, I count 10 s in my head.

End on a Good Note

Filming is the act of obtaining video footage. What you procure reflects your knowledge and efforts. You are called upon to put your back into coupling creativity, technical adjustments to your camera settings, and directions defined by your storyboard. In some sense, it is similar to a director translating a script into a play. Anyone can get his or her hands on the written document describing how the play unfolds, but it is up to you to add personality to an interpretation for interpret how it will manifest. Sometimes you will nail down a shot that is very close to what everyone else, with the same mission, would do. However, filming is a very personalized experience and most footage will reflect your own interests and tastes.

The thoughts that drive how you frame a shot and what composition you film are valuable guides for the editor days, weeks, or months down the road. So write them down. Filming is intense and exhausting. After a long day, you may feel like the day's events are indelibly stamped on your brain – never to be erased. However, this is likely not the case. A filming day should never end without a debriefing. Sit down at your kitchen table for 15 min and take notes on your efforts and your thoughts while you were filming different shots. A little bit of effort at the end of the day filming will go a long way in giving direction to later steps.

What might my notes describe? Did I film an unusual shot that might fit in where another clip was originally planned? Did I film a clip, but was unhappy with the results? Did I then have an idea on how to better use this sub par shot? Did I forget to film all the intended shots? Don't hesitate to jot down thoughts or ideas and keep your notes handy later on. You'll likely be surprised at how insightful you were at the end of your filming days.

Don't Panic

As with any major project, anxieties build and tempers flare. Since your job is to capture outstanding footage, don't panic. If something is amiss, chances are you can figure out a functional solution. Return to the first principles of filming – emphasize deliberation and quality with the shots that you do take. If you aren't sure how to film a particular scene because conditions are changing, take a moment and consider how you can use the present conditions to meet your needs or how you can manipulate your storyboard to accommodate your challenges. Sometimes changes that you make in the field enhance the way that your resulting story unfolds. Structure is great, life rarely delivers on easy days of filming. Being open to the organic nature of the filming process can deliver brilliance. More importantly, you are figuring out how to fulfill the needs of your story, despite unexpected situations.

Chapter 12
Advice on Specific Styles of Filming

Filming Interviews

Interviews are foundational for many scientific or technical films. For many people, science and math seem like very impersonal fields, for which engagement is not a comfortable experience. In fact, many people step into watching scientific films wondering why they should even start. To combat this, interviews can inject a personal feel to otherwise dry information. Showing the human face of the individuals who partake in intimidating practices, impart precious analogies, perspective, or experiences. Interviews are effective at transforming a technically based film from an inundation of information to a much larger experience in understanding how a body of knowledge is evolving, or impacting people. Interviews of non-specialists or people on the street strengthen these video productions as well. They who is impacted by a new method or new discovery. They also increase the accessibility of the concept. Most audiences, even fellow specialists, are drawn to the reactions of "real" people. Non-scientists may feel safer when people like themselves speak about information or a device that has changed the way that they go about their lives.

Filming interviews is a practice that does not come naturally to many first-time videographers. First, it is important to recognize that interviews are interactions between you (the videographer/person asking questions) and the interviewee. Indeed, you might have certain needs, hopes, or expectations for an interview, yet no matter how experienced or informed the interviewee is on a topic, they will react to and answer your questions in a manner that reflects their knowledge, personality, interests, and comfort level in front of the camera. Some folks will ramble on. Some will stammer. Others will provide invaluable insights that you never saw coming. And some of the most unassuming individuals will prove to be rock stars in front of the lens – adding personality and energy. Capturing the footage that you need, in

© Springer International Publishing AG, part of Springer Nature 2018
R. Vachon, *Science Videos*, https://doi.org/10.1007/978-3-319-69512-9_12

part, falls on the knowledge and clear words of an interviewee. Additionally, you play an enormous role in guiding or prompting them to share in a way that works best for you.

Prepping an Interview

People have varying levels of shyness. For some, being asked to jump in front of a camera for an interview is frightening. They are in the spotlight and can feel pressure to communicate clearly, especially when what they say may be imprinted indelibly on the web. Your own comfort level with conducting an interview is a very good start for building a relaxed environment. I find playful confidence to be calming. Your ease with the process can snuff out concerns that they might have about how important it is to get their words just right.

No matter whether you are asking an unknown person out of blue for an interview or you have established an email rapport, setting expectations in advance of an interview can ease anxieties. Mind you, too much information can hinder spontaneous responses from an interviewee. I try to share the basic nature of some of my questions beforehand. If the opportunity presents during the morning building up to a planned interview, I try to empathize with the person by thinking about what they might need to feel comfortable going into the interview. As such, a nurturing conversation begins long before I press record. Then, as I lure them towards the camera, I talk to them candidly about what I am doing and how I will help guide the conversation.

Here are a few suggestions of how you might approach an interviewee:

- Define the goal of your film.
- Define how interviews fit into the needs and schema of the project.
- Let the interviewee know why they are important to your task.
- Inform them on how long the interview will take.
- Avoid turning prepping into directions, where you end up telling the individual what you want them to say.

If the interviewee is creating resistance and leaning towards backing out of the interview, there are a few additional pitches that can draw them in. Comments like these have gotten me out of some binds:

- "Whatever you say can be edited. I'll make sure to use only the best morsels."
- "Would you mind giving it a try? I get that you feel hesitant, but your words would mean so much to the final product."
- "Can we start with a sit down conversation? The camera does not even need to be running." It is very important that if you take this tactic, you place the camera where you want it to be when you will later press record. Even face the camera towards the person. If that is too intense, point it only slightly away. This will hopefully warm them up to discussing the content of interest, and the feeling of a camera being over your shoulder and directed towards them.

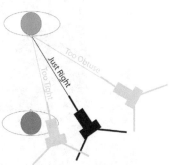

Image 12.1 Framing of an interview is best conducted with the subject staring at the interviewer just to the side of a camera. That way, the viewer sees more of one of the interviewee's cheeks than the other

Framing an Interview

During a video interview, it is useful to have a subject talk to you (the interviewer) and not the camera lens. The camera has very little personality and may remind the subject that they are explaining something for a film as opposed to being a part of a more relaxed conversation, resulting in poor-quality clips. Talking to a camera is also often not familiar to people, and this can lead to a very stiff and formal performance.

To help with this, you, the interviewer, should stand slightly to the side of the camera. As such, when an interviewee speaks to you, one of their cheeks shows more than the other (Image 12.1). Such framing gives the viewer a sense of being just off to the side of the conversation – not intruding, yet right next to the action. This effect can be difficult to achieve if the subject is far away from the camera (i.e., when the subject's entire body is within the shot). It is unsettling to view an interview where the angle difference between you, the interviewer, and the camera is indistinguishable. The small angle difference may leave a viewer wondering why the subject can't look them in the eye. Alternatively, if the subject is turned a bit too sideways, the viewer may wonder to whom the subject is talking. The first solution to this problem is to film your subject with a medium or close-up framing. However, if you choose to film from afar, I recommend having the interviewer stand in the shot with the framing capturing the side of the interviewers head and shoulders.

Note: When does a speaker look directly into a camera lens? Usually when the person is speaking directly to the viewers, like during news reports or instructional videos.

I have identified a few fundamental guidelines to help assure great framing of an interview:

Image 12.2 For interviews, follow the Rule of Thirds. This places regions of power within a shot at the intersections between the dividing lines that breaks a scene up into nine equal regions

1. Use a tripod whenever possible. However, if your camera will be handheld, keep the framing wide enough, so that any waivers won't result in the subject's face bouncing about in the shot.
2. If a subject tends towards bobbing and weaving as they speak, treat the shot the same as if a camera is not stabilized on a tripod – pull back to minimize the perceived movement and any chance that they might go off-screen.
3. Follow the *Rule of Thirds*. Visualize dividing your scene up with two evenly spaced horizontal and vertical lines (Image 12.2). This leaves nine equal-sized boxes within your viewfinder. Theory states that where these lines intersect are your power points. A viewer's eyes are drawn to these locations, making them ideal regions for interesting elements of your shot. This is true for almost all scenes (above and beyond interviews). For interviews, this usually positions the subject off of center.
4. I like to think of framing as a means of emphasizing important information. It can highlight the center of attentions. The mouth and eyes of an interviewee are the visual tools for conveying an individual's message and tone. I call these their *center of attention*. I tend to place this location near one of the intersections as defined by the Rule of Thirds. Unless you are filming an extreme close-up, avoid cutting of the interviewee's chin or forehead.
5. Most videographers would agree that leaving too much space above the subject's head and/or showing too much of their chest, perhaps down to their belt, diffuses attention on the person's face. If you tend to do this, simply move your tripod a step or two in closer to the subject (Image 12.3).

Image 12.3 Here, uncomfortable space is left above an interviewee's head. A safer bet is to locate the top of the frame such that the tip-top of the interviewee's head is as close to the top of the frame as possible without going out

Image 12.4 Positioning an interviewee's face close to the outside of the frame can lead to awkward composition

6. For interviews, artistic license in framing can distract a viewer from listening to the words. Truth be told, if you don't want the viewer to like the interviewee (or what they are saying), place the subject matter in a slightly awkward location in the shot. This conveys a slightly awkward or annoying feel that could support or detract from your intensions (Image 12.4).
7. Reduce background movements and complexity to a minimum. A viewer's eye could stray and with it, their attention.
8. Don't oversimplify the background of an interview. Bland, white walls or blue skies are boring. Include a map or some trees to break up otherwise dull composition (Image 12.5).

Image 12.5 Backgrounds that are neither too simple nor too complex complement interview shots

Image 12.6 Close-ups and extreme close-up shots excel at conveying emotions and thought

When to Use Extreme Close-Up, Close-Up, Medium, and Wide Shots

We have already talked about how to frame a person in a shot. However, let's break down how to use extreme close-up, close-up, medium, and wide shots to improve an interview.

The closer you film to the interviewee, the more expressive a person's face. If you want to show concern, surprise, anger, or confusion, fill the entire screen with their face. Eyes tell so much and the larger they are in the shot, the more gripping the clip. Extreme close-up or close-up shots may cut off part of the speaker's forehead, and the lower border is the base of their chin (Image 12.6). To be any closer is overwhelming. That said, if you want your whole film to be red-lined (i.e., every comment is important), the effect of a close-up on a face wears off. So, save the extreme close-up shots for impact statements.

Part in parcel with cutting in close to a subject is the concern of keeping them framed properly. When zoomed this close, I typically need my full attention on following the subject with my camera (on a liquid-head tripod), and as such need someone else to ask the interview questions.

A medium shot comprises the full head and upper-torso region. I find that this is my bread-and-butter interview composition. Why? Several reasons. This style of framing allows a viewer to see plenty of a subject's facial expressions and hand gestures. Additionally, I don't have to reposition my camera to track a person's movement with extreme vigilance. This opens myself up to fulfill both roles as the interviewer and videographer. What's more, my attention can focus on listening to the subject's message and for distracting background noises.

A wide shot typically includes a subject's body and then some. I think that the key to these shots is the "and then some" part. One of the only reasons to show a subject's feet (if they are standing) is for you to establish the person in an environment. Faraway shots can capture a lecturer walking about a stage without you needing to carefully track every movement with your lens. I have seen this framing applied in numerous newscasts or hosted television programs. In these cases, you are not truly embedding the viewer into an interview. This is an introduction or setting shot that might transition to a more closely framed shot. In general, I try to avoid faraway shots outside of the purpose of establishing setting.

Changing Framing

Changing framing breaks up monotony of a long scene, highlights specific information (such as close-ups emphasizing emotions or wide shots emphasizing setting), and is effective at covering up jump cuts when you cut and slice an interview down to a short length. More on this latter point later. In many cases, two well-edited minutes from an interview are more useful than the 20 min that it took to negotiate the content during the interview. At other times, an interviewee might use "umm" and the film would be better served by cutting out those pauses. Cutting and splicing a piece together from only one camera framing may leave you with perfectly flowing audio, but the video content might be loaded with ungraceful and disjointed visual cuts. Indeed, the interviewee might appear to jump from one position or expression to another. This is terribly distracting and is typically frowned upon in most videos. Earlier we mentioned how you can use B-roll to cover up these junctions, yet this would also be a great application for switching to other framing. When done artfully, the viewer may simply see the cut as a tool for changing angle, never considering that it also seconded as a cover-up.

High-resolution cameras can provide a unique method for gaining you two different framings (e.g., both medium and close-up) from one shot and thus breaking

Image 12.7 Image B is a crop in or Image A. If you can film with a camera of higher resolution than you will broadcast consider cutting in and out on a scene, in your editing studio. Doing so produces opportunities for you to cut out run-on discussions or add life to a long-running clip (*Flipping 50*)

monotony. This tool only works if you are filming at a higher resolution than you hope to broadcast. For example, you might be recording in 4K yet broadcasting your final film at 1080p. If you are in such a situation, later in the editing studio, you can cut in closer to your subject at the times when you like. This way, your single camera can seem like two cameras focused on the subject at the same orientation but different levels of zoom (Image 12.7).

Warning: Viewers will quickly pick up on posture or expression inconsistencies between angle cuts. These are dead giveaways that you are using cuts to splice together your interviewee's words. As such, be warned: audiences grow distrusting of your message if they know that you are altering the interviewee's words too much.

Image 12.8 When filming a conversation between two people with one camera, position both individuals to either side of center and ask them to speak with shoulders slightly open to the camera lens (*Flipping 50*)

Filming interviews with multiple cameras is another approach for capturing differing framings simultaneously. Effective use of multiple cameras means that you formulate a strategy for what each camera emphasizes. If I am filming one interviewee, I typically like to frame one camera (camera 1) as a close-up and the second camera (camera 2) as a medium shot. This is very similar to what I can accomplish with our earlier point about cropping in on high-resolution footage.

What if I wish to film two people conversing? If I only have one camera, the first option is to frame the scene such that both participants are framed within a single shot. Usually, I position both on either side of the frame. The closer I can frame a shot to capture both people, the more engaging the shot. What's more, if I can have a say on their body positioning, I ask each to slightly tilt their chests towards the camera, as opposed to directly at each other. This posture is more inviting to viewers (Image 12.8).

An additional way to increase the feel of interactivity between two people during an interview is to shoot over the shoulder of one subject to capture the other speaking. Most of the shot will encompass the individual who is speaking; however, off to one side, one can see how the listener's attention is drawn to the words being spoken. This is very effective in interviews. Maybe the interviewer nods their head or has an intense reaction to the words. This conjures a strong sense that the viewer is sitting in on a personal conversation. In order to keep most of the viewer's attention drawn to the interviewee, try opening the aperture on your camera (if you have the option). This reduces the focal range of the camera, drawing attention to the interviewee and providing a soft, unfocused feel to the shoulder and side of the interviewer's head (Image 12.9).

How about if you have two cameras and two participants in a conversation? How I position my cameras depends upon the group dynamic. I treat the situation

Image 12.9 Two examples of shooting conversations over one individual's shoulder. Shooting over the shoulder of one person increases the feel of interactivity between members of the conversation

differently if one person is asking questions of another than if two people are talking on equal footing. When a conversation is mostly one sided, I use camera 1 to frame the primary speaker with a medium shot and use camera 2 as a wide-angle shot that

Cool Trick #1

If you are filming a two-person conversation with two cameras, typically one camera shows the person speaking. However, breaking this rule provides a neat trick to increase viewer engagement. Once in a while, switch to footage of the listener in the conversation, while the other person is talking. Shots of a person, say, nodding in response to a statement, are often more effective than a single clip of a person rambling on for a long period of time. Additionally, this trick gives you a great chance to use cuts and splices.

embodies both. The wide-angle camera 2 shows the viewer the setting and layout of the conversation while showing both members interacting. When conversations are more balanced and I expect both people to share important thoughts, I frame both speakers with medium shots. This allows you to switch back and forth as a conversation unfolds. If I plan to move camera 2 during the interview, I try to shoot over the shoulders of both individuals whenever possible. So that I can more easily sync video from both cameras later on in the studio, I keep camera 2 rolling even while I am moving its position.

Cool Trick #2

Let's say that you have a third camera or you know for certain that camera 1 is going to cover all of your needs. Consider mobilizing your extra camera and break up any static feel to the interview. Moving the extra camera on a slider or a jib through the shot adds professionalism and dimensionality to an interview. Perhaps you will not be able to use the slider to capture every shot during the interview (or you can dedicate an assistant to do so), but an occasional, gradually sliding shot can be gripping.

Interviewee Information and Permissions

I usually start an interview by asking the subject their name, what they do, and a point of contact. You might take meticulous notes about who you are interviewing, yet capturing their verbal answers on film will serve as an invaluable backup to your memory or notes. It is critical that you remember who you filmed and how they would like to be addressed. This is particularly true if you would like to include a "lower third" (a placard, typically located in the lower part of the screen, identifying a person's name and title). Calling each person by their desired title and position is foundational to the use of people's words in films (Image 12.10).

Videographers need the permission of interviewees in order to use the footage for broadcasts. It would be terrible for you to produce a great film and one of your interviewees later protested the use of their words. These complications are avoided by having each interviewee sign release forms that give you permission to use all media captured during an interview for the purposes of your film. Very simple examples of release forms are found online; however, something as modest as this might work for you:

Sandra Larson
Principal Investigator, Strategic Toolkit University of Colorado

Image 12.10 Lower thirds are a filmmaker's go-to tool for sharing the name and position of people being interviewed (NSF, *Strategic Toolkit*)

I, _____, authorize YOU
to take video and to make use of my name for production in YOUR VIDEO'S NAME.
I give permission for YOU to edit the video content for the production and promotion of
YOUR VIDEO'S NAME for distribution throughout the world in perpetuity. I also
understand that YOU have no obligation to use the video content. I understand I am to
receive no monetary consideration for my appearance or participation in the film. I agree
that my appearance or participation in the video constitutes consideration for this
release.

_____
SIGNATURE

_____
DATE

_____
CONTACT INFORMATION

News Reports

News reports are logical and explanatory stories that typically describe one princi-ple topic or event. As with all good reporting, the goal is to share both sides of a story as opposed to highlighting one. Even though news is supposed to be revealing, information in news reports is kept simple. A major emphasis in these stories is the use of compelling imagery to bring life to information and keep the audience engaged.

Building a cohesive and fair story mandates being informed on the topic. It is then up to you to craft how you will share the information and perspective with an audience. Here are some questions that may help you move towards these goals:

- What is the story that you will cover? Is there an overarching question that you will answer? If there are background facts, get them right.
- Why should the audience care about your story? Why is it compelling, relevant, and/or timely?
- Who are the key players involved in the story, and can you nail a few representative individuals down to serve as interviewees?
- When are key elements of the story going to unfold or have they already? If they are going to happen, make sure that you are on-location at the right time to reveal the action. If they have previously happened, figure out how you can procure any previously shot media to support your story.

If a news reporting supports the message you hope to tell, find the right person to serve as the on-screen reporter. Reporters are individuals who might not have expert command of all of the information in each story, but they do have a firm grasp of the questions that need to be asked that will bring the story to life. Reporters are there to introduce the viewer to the scene and the elements of the story. As the story shifts, they will carry the viewer between locations, interviewees, or events, explaining what is transpiring or why the story has transitioned. This individual(s) will also bring closure to the story. The reporter will relate the state of affairs and how they hope or conjecture things will progress. They are friendly and composed but should not be afraid of looking straight into the camera lens as though they are directly addressing the intended viewers. In case you are into terminology, a reporter looking at the camera is called "piece to camera."

Reporters are the mortar, holding together each story. But what are the cobbles, or core elements, that a reporter introduces, summarizes or comments on?

Cobble #1: Interviews

If you are a filmmaker who is looking to use news reporting, ponder who you can interview to carry the unfolding story. Authorities, victims, or third-party viewers of an event? A scientist? A citizen on their way to work? An athlete? If you are the videographer covering the story, the responsibility is on you to take every opportunity to pull relevant people aside so that they can give you their thoughts on what happened or what is currently unfolding. Seek them out, because they will likely not come to you. It might seem like an incredible struggle at times to lock in interviews with these people, but they are powerful cobbles of many news reports.

Catching a person spontaneously for an interview takes guts and gusto. What's more, it is also challenging to know what to ask the person once you have their attention. If you are the videographer, you have a ton on your mind, and creative thinking balanced with charisma is hard to come by. What is a good place to start when conducting spontaneous interviews on a scene? "What is happening?" or "What are we looking at?" This is fairly unobtrusive to the subject and also gives

you time to measure up how they relate to the unfolding events or how their perspectives shed fascinating light on the story. What's more, open-ended questions
such as this reveal the subject's point of view without being steered.

Cobble #2: Footage

News reports greatly benefit from footage showing what happened, is happening, or
will be a part of an unfolding story. Some news reports need a camera ready to capture rapidly unravelling events. These camera shots might bob and weave – anything to catch the story! However, in general, steadier hands give much of this
imagery traction (especially true if your camera is a DSLR). As I've mentioned
many times, use a mono or tripod if possible! (Image 12.11).

Helpful Hint for Footage: Remember, news reports are meant to capture an audience. Try to think about filming elements that will strengthen the grip that your
report will have on viewers. Here are a few questions to guide you towards capturing useful footage:

- What action can you capture showing the process or unfolding of a story?
- What shots will establish the unique setting of the story? A courthouse? A
  river? If you are not permitted on premises, maybe film caution tape or "no tres-
  passing" signs.
- How do people fit into the story? Show someone looking over the rail at rapidly
  tumbling waterfalls. Capture the audience watching a talk.
- How can you juxtapose the wide shots of setting with scenes of minutia behind
  the larger story? A gavel (even a still shot) in a court case. A lone raindrop drip-
  ping off a leaf during a receding flood.

Cobble #3: Green Screens

Let's say that you are the reporter and you want to explain a concept. Some ideas are
hard to weave together with only words or simple B-roll shots. Green screens can
help you out. Green screens are tools for creating backgrounds to scenes that might
not be humanly possible without this technology. Perhaps you want to be a news
reporter standing on the moon or next to an animated weather map. Green screens
allow for these options. It begins by simply unrolling a giant tapestry of bright blue
or green material and hanging it behind the reporter. You can also paint a wall with
these colors. When you film a reporter standing in front of the screen, the bright
green will stand out as an obnoxious background. Once in the editing studio you,
the editor, can choose to replace the green color with another picture or video clip.
If you have footage of clouds, that can be your background! If you want an animation of satellite imagery to unravel off to your side, you can make that happen! What

Image 12.11 Two examples of compelling supporting footage, used to enrich news reports

if you don't have a newsroom but want to reproduce one? Use a green screen! Magic manifests with green screens. Not all editing programs give you this option, but many do (Image 12.12).

Note: Green screens are useful for applications above and beyond news reporting. They are fantastic for unraveling science fiction tales where you might want explorers walking on a faraway planet. Alternatively, you can have an individual indicating different steps in a manufacturing line, as represented with video footage in quadrants surrounding the speaker. The uses of green screens are limitless and are particularly helpful during news reports.

Image 12.12 Green screens are a simple yet incredibly powerful tool changing the background of a news report. First, film with a reporter positioned in front of a green screen. Later, while editing, replace the green color with a more suitable background. Figure out whether your editing package allows you this option

Cobble #4: Storytelling Guide

News reports often follow fairly simple master templates for how stories are approached. They follow structured pilots such that numerous reports can follow similar story arcs. A series of reports might begin with a very short (a few seconds) introductory animation ending with the program's title. Then your news reporter will step in, introducing the story and the relevance of what is going to be covered. Next cut to the story on a certain location. Maybe the reporter is speaking while illustrative B-roll describes the situation. Or you cut to a sequence of interviews held together with connective perspectives from the reporter. Conclude each story with your core points, usually told through media that is supported but not driven by the reporter. Lastly, the reporter puts a bow on the package – condensing the message and sharing parting thoughts on what happened or what might happen next, keeping the viewers coming back for the next news report.

Cobble #5: Teleprompters

Some of the best reporters speak eloquently about topics without a script, but that is not always the case. If you are new to the role of being a reporter or you would like the news reporter to use specific well-phrased sentences, try using a teleprompter. A teleprompter is a tool that scrolls the words that the person hopes to speak. The hope is to position the text close enough to the camera such that viewers cannot tell that the individual is reading. Higher end teleprompters reflect text in front of your camera lens. That way a reporter can look into the camera lens and simultaneously read a script. Special clear glass scrolls words while letting the camera "see" through the words to capture the reporter. A shroud surrounds the reflective material to prevent unwanted light from hitting the screen. Since the teleprompter screen is not very large, the video camera and thus a viewer typically do not pick up on the speaker's eyes as they follow along with the words. Unfortunately, if the videographer zooms in too closely to the reporter, eye movement can be more easily made out. This usually does not go-over well with viewers, so avoid close-ups.

Teleprompters, as described above, are intricate. Professional-grade devices cost in excess of $1000. This is a pretty penny; however, they have improved production of many television shows and as such prove to be great investments. That said, there are some less expensive options for devices that can fulfill the same needs. One can download applications for your computer or tablet to serve these purposes. Many scroll text, just like a teleprompter; however, the words display on your monitor rather than a reflective screen. What is the best way to position your screen? Nuzzle it up very close to your camera. In fact, professional-grade teleprompters normally project the scrolling words slightly below the center of the camera lens. The challenge is figuring out how you can fashion the screen of a tablet directly under the lens (as close as possible). A simple music stand may serve these needs (Image 12.13).

For successful implementation, it is helpful to have a dedicated assistant assigned to the teleprompter – watching the monitor, pacing the scrolling words, and moving from one text file to the next. What's more, if mistakes in spoken word are made, the assistant will have to move back in the text such that it can be replayed. These are minor complications but can slow production.

Digital Animation

Many of us are familiar with digital animations. We see animated titles to movies or corporate logos swirl in front of us. What's more, computer-generated imagery (CGI) has swept the world with iconic movies such as *Shrek* or *Zootopia*. Feature-length movies of this nature are incredibly work intensive, but digital animations on a simpler scale can also be very compelling. Indeed, the process of making digitally animated videos produces new film content; however, we will reserve discussions about the fundamentals of animation for the chapter on movie editing. Instead, here we will focus on a couple other animation techniques that require camera usage.

Image 12.13
Teleprompters can be
approximated using
well-placed tablets or
computers. Once outfitted
with an appropriate app,
one can program a script to
scroll at the pace that suites
a reading/speaking speed

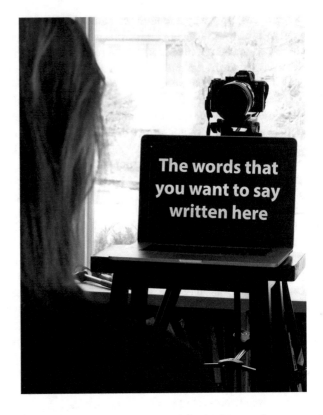

Sharing Abstract or Complicated Concepts with Animated Drawings

Sometimes literal representations of settings, processes, or conceptual relationships are not easy to film. Let's say that you want to show how certain animals come out at night. First, it is hard to film in the dark and nocturnal animals are often shy. Sure, you could spend quite a bit of time and effort to film this, but you can also use animated drawings to more quickly and effectively show what creatures might be doing.

Do you or someone that you know have drawing skills? Not only can drawings be introduced as individual still shots, but filmmakers have also found great success capturing the act of scrawling a drawing and embedding sped up versions of this process within movies. Here we will talk about the basics to get you off the ground.

First, find a table with plenty of diffuse light shining down from a number of lights from above. Place a clean piece of white paper on the table and tape it into place (so that the paper does not wander as you draw). I like to use paper that is quite a bit larger than 8 ½ by 11, so that you can draw a large, detailed image. Even use a dry erase board! Now set up your camera and tripod to shoot over your shoulder, making certain that your body does not throw strange shadows across the drawing

Image 12.14 One of the simplest form of animated drawings is to film yourself drawing a picture. Later, while editing, speed up the process to last a much shorter period of time during. A drawing that took 15 minutes to draw could explain a concept in only a couple. Many complex concepts have been explained efficiently using this method

surface. If shadows are unavoidable, try to keep them from falling over your pen or where your drawing will appear (Image 12.14).

In these situations, it is very important for you to know the framing of your camera. In general, I keep the edges of the paper or dry erase board from being seen, such that the viewer doesn't see odd boundaries or tape. I make certain that all elements that I draw are visible. I often put light marks at the fringes of my camera framing to let me, the artist, know what boundaries I cannot pass with my pen.

Now it is time to draw! You may be tempted to jump straight to pressing record and start drawing. However, practicing the drawing of your image hones in on how you will portray the objects within your image and the sequence of which elements will be drawn first. Indeed, the timing of when you will draft each element will carry your viewer through the story that you tell with your artwork. Don't draw a cat first, if it is going to chase a dog. Draw the dogs and then bring in the cat as the attacker! If need be, write directions on a separate piece of paper that you can refer to just out of the framing of your camera. Lastly, practice drawing. This exercise keeps you from pulling your hand away from the paper, or bobbing and weaving as you consider where you will draw your next element. These oddities distract from an unfolding story. That said, pauses during your animation give you time for extended discussion about what has transpired at that certain point in your animation.

Recording these shots takes some preparation. Before drawing what you hope will be your "keeper" clip, place one of your practice drawings on the table. Take a moment to adjust your camera's light settings such that the paper color is white and your lines and colors vibrant. You will probably note that strong colors, like navy blue, black, and red, show up best, especially when you draw with a thick felt pen. What's more, your hand is not the most important piece of information, but your camera does not know this. This means that you have to guide your camera to adjust for light and focus upon the right information. Use plenty of light! With lots of light, you can close down the aperture on your camera, thereby increasing your range of focus (so that the paper and your hand are crisp). Speaking of focus, recording with autofocus can be disastrous. Hand movements may draw focus off the paper. Instead, manually focus on the paper.

Once you press record, take your time getting in position. In the case of drawing on a table, I have found that in order to have the camera positioned over my shoulder, the tripod is often right beside the back of my chair or on top of the table. I must use caution, as some movements might bump the tripod and ruin smooth recording. Sit comfortably. If possible, have someone observe the viewing screen and tell you where you can position your head so as not to block the camera lens. The same person can watch through the camera as you draw, making sure that your body stays in an appropriate position.

Now that you have captured your clip, import it into your editing program. Showing clips of drawing, without speeding them up, is often tedious for viewers (not always the case, especially if you have the right narration to accompany the drawing). Speeding up by double can come off a little strange. I am a big fan of speeding this process up by a factor of three or more. Anything less has an air of unprofessionalism. Lastly, try including the sounds from the drawing process. Scribbling, even in high speed, can give a signature sound that gives these scenes greater impact. Alternatively, eliminate the sound of drawing and add fun music!

Stop Motion Animations

Animation is a very difficult undertaking in general. The learning curve is quite steep and long. Some would say that even more taxing than traditional animation is stop motion animation. What is stop motion animation? Stop motion animation is the stringing together of several photographs that capture a setting whereby each subsequent image embodies only slightly different positioning of props, such as *Legos* or *Play-Doh*, within the shot. Typically, a camera is steadied on a tripod while you manipulate the props within the frame of your shot between photographs. With images played sequentially, the props take on movement. Masterpieces of stop motion animation include the *Legos* movies or *Wallace and Gromit* (Image 12.15).

Stop motion animations are stylistic and ground a viewer to tangible moving parts. As such, the concrete medium for your props and characters may be more forceful than drawings or computer animations. What's more, creating seemingly natural activity with objects that people otherwise know as stationary is exciting. Some people have cited stop motion as a method that throws back to decades ago and can give an old-fashioned feel. This can either work for or against your greater goal.

Image 12.15 Stop motion animations use photographs of gradually manipulated props to constitute the individual frames of film clip. Played as one in a VEP, the progression of images will present movement. Here Image **a** transitions a viewer from a Lego horse's face to that of the Lego rider in Image **b**

Most stop motion filmmakers agree that time is one of the most critical factors to consider when undertaking a stop motion project. The gradual, incremental, and natural movement of props within a scene requires creativity and vision, and the process is painstaking. As such, stop motion animations are usually captured with the use of small models of reality. However scale is not a defining quality of stop motion animation.

The production of stop motion film also requires meticulous storyboarding. Why might storyboarding be more important with stop motion than other forms of film? Re-filming a scene is an arduous and frustrating task as it may require the re-articulation of thousands of movements. What's more, if you forget to include the action of one element in a scene, even for a couple of clips, it is very hard to go back in time and return elements to the positions that they once were in. The best prevention methods for these challenges are careful planning and vigilance to following every step within your storyboard.

As with any mission, knowing the complexities of your task, with all of its idiosyncrasies and pitfalls, is the name of the game. Here is a collection of pearls of wisdom that can save you time and effort as you learn to make stop motion films. Much of what I share below comes from personal experience, conversations.

These will be helpful, but the clincher steps to making good stop motion videos is starting small, making lots of mistakes, and learning from those mistakes. You will make enormous gains between your first few efforts. Allow yourself time to make those leaps before you jump headlong into a very important project:

- **Pearl of Wisdom #1**: Many still shots linked together within every second of footage will lead to smooth progression of movement within your scenes. This is true for all filming. If you can match the frames per second of standard film rates, 30 or 24 fps, your videos will appear incredibly professional. That said, to capture upwards of two dozen photos for every second of video content is an arduous task. While the high frame rate optimizes the fluidity of a clip, shooting 12–15 separate images per second of content provides equally exciting results. Anything less than 10 fps results in mixed outcomes.
- **Pearl of Wisdom #2**: Use well-designed shortcuts. As with all videography, the audience will likely never know the measures that you take to film a scene. Those are locked in the darkness of what happens behind the camera. So, take time to think of simple ways to reduce your labor as you film a stop motion scene. Simplification does not mean cutting corners. What do I mean?

  - Use puppets instead of manipulating models. Hands, along with rapidly firing cameras, can cover immense ground. So does pressing record on a video camera and later manipulating the footage to appear like a scene was shot in stop motion. You can always slow down or coarsen a scene shot at 30 fps (standard) to, say, 15, thereby matching frame rates with ensuing truly stop motion-created clips that were filmed at about 15 frames per second.
  - Duplicate photos when you want to slow down a scene. If you find that your scene unfolds too rapidly in places (perhaps if a character is stunned by a thought), copy and paste two of the exact same photographs together, thereby creating a pause.
  - Film a smaller scene than you might originally imagine. I am not talking about the size of your stage. Larger and complicated sets are difficult to change between shots. They are fantastic for establishing a setting, but close-up shots may include fewer moving parts.

- **Pearl of Wisdom #3**: Lighting. Even though you might not notice that the lighting in your scene is changing over the course of shooting session, it oftentimes is. This can easily happen if you are relying on natural light from windows to illuminate your set. Clouds or shifting sun angle produce variable lighting that can ruin what you hope to be a clincher scene. Instead of leaning on natural lighting, use light from a number of controlled sources. Try to avoid too much direct light to reduce high contrast and deep shadows. For example, *Legos* in stop motion animation have a tendency to gleam in direct light, which is distracting. A simple fix to this conundrum is to place wax paper in front of any high-intensity light bulbs. Throughout a shoot, avoid moving your light sources unless the shifting of light is the effect for which you are aiming (replicating, say, a sunset).

- **Pearl of Wisdom #4**: Once you have set your stage it may be tempting to jump headlong into filming. Rushing is often a terrible mistake in stop motion productions. Take extra time to make certain that you fix your camera into the position where it will remain throughout a scene. This can mean duct taping your tripod legs into place and locking-down all adjuster knobs. Why? Let's say that you want to film a 5-s scene and you plan on staging about 12 shots per second in your final film. Over the course of manipulating your elements 60 times to produce a full scene, there is a good chance, unless you are very careful, that you will bump the camera. Even if you nudge it a wee bit, you will either have to take extreme measures to realign your camera, deal with a less than perfect clip, or reshoot the entire scene.
- **Pearl of Wisdom #5**: In keeping with the thread of locking in your tripod and camera, use of a remote camera trigger helps. A remote trigger allows you to take a photo without touching your camera. Even though you might think that pressing the trigger on a camera does very little to upset the framing of a shot, it does. This is especially true if you are using an inexpensive tripod, framed at a strange angle, or zoomed in on your set. Remote triggers, which can be purchased separately for some cameras or can be downloaded as an app for your smartphone, make for liquid smooth transitions between individually shot frames.
- **Pearl of Wisdom #6**: Simplify how you fix your scene props into place. Between just about every shot, you will be putting your hands into your set to manipulate characters or objects. While your hands will be the devices to change the position of these props, you can overdo movement or knock elements over. Imagine trying to make a character walk across a street and one position knocks the character off balance. How do you get the character back to the exact right position? As such, use elements that will hold positions and lock into locations. This is why many people producing stop motion films use *Play-Doh* or *Legos*. It is suggested that if you use *Legos*, you use a *Lego* base plate as your ground. What's more, use double-sided tape under these plates to keep them from sliding on your set table.

Once you have secured your shots, it is up to you to link them together with video editing software. In these programs, you can compile the sequence of images and export them to a video file that all can view. Old-school filmmakers used to take individual images and insert them into an editing program, one at a time. They would place one shot after the other and trim them down so that each shows for the same tiny amount of time. Modern editing gives you much more efficient tools for the task. As long as your photos are titled numerically, editing programs dedicated to building stop motion videos will take care of these needs automatically. What's more, you can define the amount of time that a scene will occupy. Let's say that you want to use 15 fps to show a certain scene; however, after the images are assembled, you might decide that 12 fps is more natural. Modern editing programs make these changes in a few short seconds.

Video editing software for your computer is not the only thing to have made enormous leaps in technology. You can download free or inexpensive apps from the web that link the camera on your phone to simple on-board editors. Hold your camera in place. Shoot a few shots that you intend to place in each intended second of production. The app will piece all photos together at your desired speed. What's more, you can add voices and sound effects quite easily, even while riding in a taxi to the airport. Stop motion filmmaking has never been simpler!

Personal Anecdote

Much of my experience with cameras burgeoned while I was a youth taking loads of photos in wild parts of the United States. During my college years, my buddy and I took half a year off from studies and traveled around the national parks. We lived off of a shoestring budget and puttered from place to place with a $700 car. He had his guitar and I had an aged, 35 mm film camera. He would sit at camp strumming for hours and I would trot into the wild photographing the bounty of American landscapes – Zion, the Tetons, Yosemite, or Big Bend.

Each roll of film allotted me 36 photos (38 if I was lucky). Shot preview was limited to the viewfinder. I could not take a shot, review it, think about how it could be better composed, and then retake the shot a moment later. I was motivated to improve my photographs by hope, anticipation, and accumulated experience. I would try a number of light settings and framings of composition, knowing that one would be better than the others. When we emerged from a week in one park, my buddy would replace a guitar string or find sheet music, and I would dash to the local photo store to develop my film. I relished getting my film back – seeing how my skills for approximating what would make for better shots improved. I think that enthusiasm drove me to appoach growth systematically. At first, I'd try adjusting one setting and seeing the effects. I would then manipulate them in tandem. This informal time spent in the wilds with my camera slowly, but surely, honed my eye for composition, lighting, and framing (Image 12.16).

Landscape photography and videography skills take time to develop. Sure, you can read a book or talk to friends. From there you can create your own opinions about other people's shots. It is a different story to be able to manifest great shots on your own. Whenever possible, get out, have fun, and learn your camera and how to compose great shots. Experience directly feeds back on consistently delivering on quality video clips.

Image 12.16 Photograph of a Northeast seascape

Landscape Shots

Many people know the concept of "landscape" because of printer settings. Through this context, a landscape shot is the opposite of a portrait – meaning that the longest side of your framing is oriented horizontally. In truth, "landscape" comes from the paintings and photographs comprising mountains, towns, farmland, etc., where the framing best fits the content when the canvas or camera is oriented with the long side fashioned horizontally. Since video is almost universally filmed in this orientation, our definition of the word *landscape* will hinge on the content within the scenes, such as cities, pastures, or mountains – generally wide-open spaces.

Filming these scenes is more complicated than fixing your camera to a tripod, relying on the automatic setting for light metering, focusing, and pressing record. This is a good start, but you can make several adjustments to enhance the impact of landscape shots.

Even though experience pays dividends for shooting great landscape shots, there are some practices that give your work a healthy boost. Here are some staple techniques that most landscape videographers apply to most situations:

- A great starting place is to understand how you can manually adjust exposure compensation, ISO, aperture, and shutter speed to change the ways that light is characterized in your shot. Refer to the previous chapter.

Image 12.17 Film such that the most important elements in your scene come to life. Here we wished to draw eyes towards the explorer

- Use a tripod. The mission of filming landscapes is to tell viewers where a situation is unfolding, and any shakes or bumps to the camera distract a viewer soaking up all of the details of your composition.
- Think about the purpose of your shots, and cater your framing and composition towards these greater goals. In general, a viewer likes to have their eyes drawn towards something grounding in a shot: this can be an object(s) or interactivity between objects. Without something to focus upon, minds wander. Consider – is the broader landscape what you want to highlight or do you wish to illuminate how smaller elements fit into the scene? Are you filming to describe active relationships, like salmon hurdling up rapids, or more stationary positioning of elements within a setting, like a hunter in an expansive wilderness? (Image 12.17)
- Consider closing your depth of field to include elements only in the foreground or the background, yet not both at the same time. When both are in focus, a viewer's eyes may be distracted from an animal when it is perched in a setting. Blur out the landscape a little bit and the creature pops out!
- To build upon the last comment, try a *rack focus*- showing how foreground elements relate to a setting. For example, you may adjust focus from a pigeon on a rooftop to the city surrounding it. Now the viewer knows that you wanted them to see the bird and intuitively follow the transition to thoughts about the city. In order for this exercise to work, one must be vigilant not to shake the camera while the focal range is being adjusted. We discussed this technique as a cool trick in Chap. 11.
- Some videographers lean on the *Rule of Thirds*. We mentioned this rule when we discussed how to frame interviews, yet it also applies to landscapes. Use the

Image 12.18 Landscape shots benefit from holding true to the *Rule of Thirds*

intersection of the dividing lines to draw the viewer's eyes to these exciting locations of your scene more easily (Image 12.18)

Lines draw attention to key focal points. For better or for worse, lines that are strikingly different than the rest of a scene draw in a viewer's attention. Why? It has been cited that the mind turns to wondering what is the meaning anoolistic geometry and is curious about its nature. A great example would be power lines cutting across a valley cloaked in mist and autumn colors. It shatters the serene landscape. Is that what you want? In some cases, no. In others, yes! What if your story revolves around the negative impacts of a local power plant? The power lines will convey the ruination of something beautiful. What's more, you can depend upon audiences to fabricate their own lines within a scene. This concept is best illustrated with examples. When composed correctly, people will want to know what is around a corner in the road. Additionally, viewers want to know what has captivated characters within a scene, so they will follow lines of sight until they determine the point of attention. If the object is critical to your story, you have created a very powerful storytelling tool (Image 12.19):

- The sky is a critical part of many landscapes. If your sky is dramatic, consider letting it occupy the upper two-thirds of your scene. If it is more ho-hum, tilt the camera to occupy more of the land.
- Adjusting a camera so that both the sky and landscape are well lit in your scene presents challenges for all videographers. Skies during the day, particularly at sunrise/sunset can leave the land deep in shadows. At times it is next to impossible to capture a well-detailed sky and land at the same time. What can you do? One solution is to wait for the hours when the sun is lowest yet still making

Image 12.19 (**a**) The use of lines can carry a viewer down the road of your story. (**b**) Lines are not always literal. Viewers are drawn towards following other people's line of sight

contact with the ground. Alternatively, let the landscape fall into a dark silhouette. Just make sure that the silhouette is interesting. Another solution? Looking towards a setting sun can deliver great shots, but try turning 180 degrees around. Incredible colors might be happening just behind your back! A third and more complicated solution entails the use of filters that you can buy from photography stores. Some are designed to darken the sky and draw out the less well-lit land below. Using filters is an advanced technique that takes time to learn and apply (Image 12.20).

- Speaking of filters, consider using a polarizing filter. When light is bouncing off of objects, like lakes, polarizing filters take away some of the glare. At the same time, these filters extract more blues and greens in other parts of the scene. The results are profound.

Image 12.20 Sky can be the strongest element in a scene

- Listen to nature. Landscape photography and videography benefit from paying attention to what nature is doing and adjusting your camera to let those qualities come to life. A very simple example includes the harsh and direct sunlight that bathes many landscapes during the afternoon. Understand this dynamic and work with it. If you want to film a car blasting down a highway bathed in heat waves, this time of day might be ideal. However, it is less ideal for other situations. Ponder when sunlight will circle around a mountain and hit the mist of a waterfall, letting its grandeur unfold on your screen. Design your day around nabbing the best shot possible. Oppositely, be open to letting go of your plan to leverage nature's vagaries to your advantage. For example, what if storm clouds brew most afternoons, occluding views of where a city fits into its surrounding? You might think that a setting shot will have to wait for a clear day, but be aware – the most spectacular shots happen when weather transitions. If there is a chance that the clouds will break, watch what is unfolding in front of you. You might capture mind-blowing footage that you otherwise would not have foreseen, like a break in the weather manifesting a rainbow! Unfortunately, being open to nature's impulsive beauty means that you need to be mobile and more flexible with your time (Image 12.21).

Time lapses may bring landscape scenes to life. Landscapes are typically shot from afar, rendering scenes simple, expansive, and occupied by few moving elements. Such scenes show little activity and interaction, which might not support the nature of your story. What if you were producing a film about the shipping industry and have the chance to film a harbor shot from the 50th floor of a building. Filmed at normal speed, freighters may appear as distant stationary dots. A simple solution to this problem is to record the scene for 10 min and later speed it up, in the editing studio, to occupy 4 s. The lumbering freighters might materialize.

Image 12.21 Two examples of how waiting waiting for the right moment can produce mind-blowing results

Screenplays

Screenplays are written documents that describe how a scripted movie or show is supposed to unfold. The job of a screenplay is to inform all of the people involved in filming about the vision of how the writers of the film want a cinematographic work to sound and appear. They include written portrayals of transitions into and out of each scene, movements of the camera, actions and expressions of the actors, sounds, and, of course, dialogue between characters.

It benefits someone who wishes to write a screenplay to align his or her efforts with the established screenplay format. Indeed, there is convention. It may seem like there should be more than one way to execute this activity; however, one format over others is your pathway into pitching your work to other film producers (should you be interested in that side of the industry). Many producers expect adherence to this format.

The details of how you should write a screenplay can seem overwhelming, and overly specific; however, with a little time in the saddle, the formatting requirements become second nature. Where do you start? Here are a few basic rules and sample is shared below. Personal notes about the nature of screenplays are included parenthetically (Image 12.22):

- Each page must contain all of the descriptors that will fill 1 min of playing time. This is an approximation, and the final product is often different from what you expect.
- All text should be formatted to *Courier 12.*
- Pay attention to follow to document layouts, which means tab settings and capitalization assigned to scene headings, transitions, dialogue…. For example, your slug lines must have 1.5″ margins on the left and 1″ on the right.
- It is generally accepted that, at the top of each page, you write "FADE IN." What's more, the end of a page usually includes "FADE TO BLACK." More details or other transitions can be included; however, without a greater understanding of your scene, this is what is expected.
- How you describe the use of technical elements in a scene follow a strict code. Technical elements include sound effects, camera angles, scene transitions, speaking from outside the camera's frame of reference….

The standardization of the formatting is a formality of the practice; however, the methods are effective at embodying all necessary elements that producers need. If a screenplay is what you wish to write, I would suggest spending a lot more time reading an entire book devoted to writing such manuscripts. What's more, check with your intended recipients of the screenplay to understand their specific needs for screenplays. What's more, have a look at sample screenplays online.

Final Thoughts on Video Production

Production is a very broad term for how you acquire the media that will be included in your film. Here, we discussed several methods to help you begin to grasp how to approach particular styles of production. The purpose was to illustrate how the fundamentals of filming, coupled with focused intent, can manifest very specific and exciting outcomes. It is empowering to know that you can focus your efforts and tools for faster and more tractive production. As such, before enterprising into a production of a certain style, take a step back and conduct up-front research on the options of film styles that might interest you. Talk to people who have filmed with methods of interest. Watch other videos that use these techniques and see how they elicit the tones and convey information that captivates your attention. Learn from others so that you can use their practices to convey your story. These steps, done in advance, will help streamline the production process.

FADE IN: (the very first entry on the first page must be this
[all in caps])

INT. COFFEE SHOP - 9 IN THE MORNING (the when and where of a
scene - EXT refers to exterior and INT refers to interior [all in
caps])

BARISTA, a 22 year old woman who clearly did not get any sleep
the night before, stares blankly at the hissing espresso maker.
(author's description of what is happening in the scene, and
character names or titles are in full caps)

JOHN, 45, hardly needing a coffee at all, pops up to the bar.

                    JOHN (2.2" over from the left margin)
          I'll have exactly what you think
          that I should have on this fine
          morning. (1" to the right of the
          left margin and 2" to the left
          of the right margin)

                    BARISTA
          How about a cup of go away. I
          can't imagine that anything that
          I can do will add happiness to
          our planet.

                              FADE TO BLACK (optional
                              scene transition)

FROM BLACK (optional)

EXT. BUILDING ACROSS THE STREET - SAME TIME

HARRY THE BIRD, a disheveled bird, hobbles across the sidewalk to
stand on the edge of the curb. Smell the air as though it was a
rose. Lifting his wing, like a hand puppet, he mutters.

                    HARRY THE BIRD
          I can't blame you for hanging
          out with me again today. I can
          really serve up a mean city
          experience.

                    HARRY'S WING-PUPPET
          You got it boss. I am always
          here for you. What I need right
          now is a good stretch and...
                    (MORE)

Image 12.22 Example of screenplay structure

Chapter 13
Recording Sound

The Math of Sound

Before we step into the methods, there is a critical concept of physics that might help guide how you use microphones. Mathematically speaking, sound level drops by the cube of the distance from its source. This is easier to intuit if you think of sound being density. Sound at a source is very dense, yet as it moves out in a sphere, volume increases as a cubic function. So the sound at the source that was dense must then spread out to fill in the larger volume. In simple terms – moving a camera in, closer to the sound source, rapidly increases the chances that the sound that you are recording is dominated by your chosen microphone! (Image 13.1).

Monitoring Sound Quality

No matter where or what you are filming, steps must be taken to assure that quality sound is being recorded. While certain measures increase your chances of producing a good soundtrack, it is useful to verify that you have captured the best sound possible while you are actually recording. Remove the chance of subpar soundtracks by monitoring the sound being recorded, in real time. Listening in on the sound that comes from your microphone(s) gives you the power to observe whether the methods that you are using are working. If they aren't, you can make on-the-fly adjustments accordingly.

Most prosumer cameras or portable sound recorders come equipped with an earphone jack. This plug taps straight into the sound that your microphone is capturing. There are only benefits for using this tool. At a first approximation, you can hear whether sounds are coming in too loud or too soft (some cameras have bars showing this information as well). You might hear the jingle of a necklace during an interview. Or, it will inform you about whether your body position is buffeting the

Image 13.1 Sound decreases very rapidly with distance away from the sound source. The relationship between sound level and distance from the sound source is an inverse cube function. Because of this rapid drop in sound volume, locating a microphone near to the source can be an effective step towards reducing background sounds

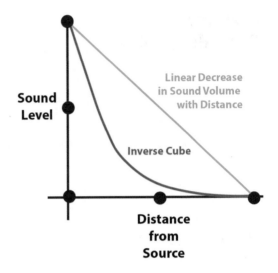

shotgun microphone from high winds. Maybe you will learn that you did not even turn on the power to your lapel microphone transmitter!

Cool Trick
Some interviewees become distracted by videographers/interviewers who are wearing earbuds to monitor sound while conducting the interview. Some cite a feeling of detachment. I have found great results from using one earbud in one ear and directing my open ear towards the interviewee. If this act is done subtly, they may never even know that you are monitoring their sound through the camera.

Always take time to assure that the sound that you want is the sound that is being recorded.

From an application perspective, different audio recording devices excel in different scenarios. Whether you have any number of microphone types at your fingertips, or you know that most of your work will revolve around certain uses, having the right microphone will save you time and energy. It will also likely result in a much better sound. Earlier, the types of microphones that you can purchase were touched upon. In this chapter, the discussion will be guided by the possible sound situations that might present and which microphones will serve you best during these instances.

Capturing Background Sound

Whether you are filming in a forest, in a museum, or at a family dinner, there is great value to recording the sounds of the environment. Although some viewers of films may not recognize how ambient sound helps to place them in a visual setting, sound is very effective at embedding a viewer into a scenario. When on a shoot make sure to record at least 10 s of background sound at each stop. In general, I like to record some background sound without being too selective about from which sources it is coming. This means that I don't emphasize a specific element, like a singular car in a traffic jam. For these purposes of catching holistic background sound, cardioid microphones excel. Most microphones that come standard with cameras are cardioids (however, some high-end camcorders come standard with shotgun microphones). The cameras with standard cardioid microphones do a decent job of capturing omnidirectional background sound. Even better, use microphones dedicated to this sort of omnidirectional sound capture, such as a portable audio recorder.

For capturing ambient sound from one source while blocking out others, a shotgun microphone is best. For example, if you are in the woods and you want to capture chirping spring peepers yet there is a nearby highway, you can use the powers of a shotgun microphone to reduce the road effects. You would execute this by positioning yourself such that two requirements are met: (1) the shotgun microphone is pointed towards the spring peepers and (2) the road noise emanates from beside the microphone. Doing so will minimize the road noises in your recording.

This brings up a very interesting point. You may not be able to position your camera to visually frame the angle that you want to film, while simultaneously capturing the sound that is optimal. Don't let this stop you from recording plenty of optimized sound or filming the video that you like. Why? Because once you are back in the studio, you can use the great video clip, while overlaying the good audio track over the suboptimal audio track. This way, viewers will see and hear just what you want. The use of replacement sounds (like in this case) to fill in for the sounds that you captured while you were recording video is called *foley*.

Note If the only camera that you have is a sports camera, like a GoPro, make sure to remove any waterproof cowlings. While protective against water, they ruin sound recordings that are already lacking in most sport cameras.

Recording Interviews

Many films are bolstered by interviews. By definition, you are using the words of individuals to tell part of your story. Just as though you were watching a film where a speaker's face is out of focus, unclear voice recordings can limit film impact. This means that you, the videographer, should make every effort to remove sounds emanating from sources that are not the speaker's voice.

The first step in this process is to try to eliminate the background sounds. Seek out a quiet room or wait patiently for a train to go by. If you are in a room with other people talking, find the most quiet corner or request that people in nearby conversations keep their voices down. Also, be aware of less obvious sounds that you might not recognize because they are constant. For example, the gentle and constant whirring sounds of ventilation fans or far-off traffic can detract from a good recording. In advance of any recording where sound is paramount, take 10 s to listen to what your camera (or recorder) is capturing. This might be with a sample recording or simply listening in with earbuds. Slow down and listen attentively to the qualities of surrounding sound. If you detect any distracting noises take as much action as you can to optimize. Also, pat yourself on the back, because you caught the poor-quality sound before you started your interview.

The most efficient and effective ways to capture voices depend upon your relationship to the interviewee. Should you know the individuals well or have organized an interview long in advance, most videographers use lapel microphones for this job. Generally, these little microphones are clipped onto the buttons of a shirt in the middle or upper middle segment of a person's sternum. At the same time, the location is close enough to the speaker's mouth so that their voice will be much louder than most background sounds. One might think that a microphone in such a location would prove to be a visual distraction; however most lapel microphones are small enough to remain fairly innocuous. Also, the cable that leads to a transmitter or to the faraway camera can be tucked in a person's shirt, thus remaining hidden.

Positioning a microphone in such places puts it in the potential line of fire of heavy breathing. If an interviewee has slumping posture that often happens in a lounge chair or they laugh while breathing downward, be wary that their breath could create disturbances in your recording. Placing the microphone slightly off to the side can resolve this issue.

Opposite to having preparation time before an interview or having familiarity with the interviewee, what do you do when you have the opportunity to interview a person on the fly? There is a good chance that you can still offer to place a lapel microphone on the shirt; however this step is slightly invasive and takes time. Instead, turn to a shotgun microphone.

When working alone, one can position the shotgun microphone onto the hot shoe of your camera body. In this location, the microphone faces forward and subsequently wherever the lens points, the microphone will pick up the sound from the subject you are filming. With an interview being recorded, distracting sounds to the side are minimized and you can typically expect decent voice recording. This is particularly true the closer that the microphone is to the subject (Image 13.2).

Be aware, longer shotgun microphones might stick over the lens. For some wide-angle shots, this could interfere with video framing. I have had the greatest results from medium-length shotgun microphones (Image 13.3).

Shotgun microphones can also be placed at the end of a boom pole. This gives the handler of the boom great flexibility with the location of the microphone.

Image 13.2 When mounted to the hot shoe of your camera and directed straight forward, shotgun microphones excel at capturing sound that originates from the direction that the lens is pointing

Image 13.3 Shotgun microphones can be purchased at different lengths. While longer microphones can focus on certain sounds more accurately, when positioned on the hot shoe of your camera, they may intrude into wide-angle shots

Image 13.4 There are numerous ways to fix a shotgun microphone on the end of a boom pole. This gives you great freedom to transport a microphone to great heights or to regions just out of view of the camera. Booms fascilitate unexpectedly great sound capture. Here we were able to construct a $30 boom with painting accessories

Typically, with the help of an assistant or other videographer, the boom pole is positioned either above or below the interviewee, just out of frame. Placement requires more thought than simply maintaining a close position to the subject's mouth. The person in charge of the boom must direct the barrel of the microphone towards the sound source. It is a true craft to direct the microphone and know if the microphone on a boom intrudes into your shot. While videographers should inform you about such interruptions, I think that it is best to work conservatively and place the microphone a little further out of the shot than you might think. With time and careful communication with the videographer running the camera, methods are easily dialed (Image 13.4).

Note It is possible for one videographer to run the camera, ask the questions, and direct a boom microphone effectively, but experience has shown that learning this craft takes practice. You also still run a higher risk of disappointing results.

What if the unfortunate scenario were to unfold where you want to record an interview, but you showed up with only your camera and a portable sound recorder. You are presented with the conundrum of how to capture great video and audio, but don't have a lapel or shotgun microphone. I would suggest framing your camera perfectly and arranging the audio (on your camera) to whatever setting best catches the person's voice. Don't worry, sound recording on your camera does not have to be perfect, but you have to be able to distinguish words later. Then take your audio recorder and place it on the table just in front of the interviewee, keeping it just out of the camera's framing. Press record on both. Later, overlay the good sound, captured by the recorder, over the poor-quality track captured by the camera.

Cool Trick

Returning to the concept of foley (using sound captured from one location (or scenario to provide useful audio for a video clip captured at another place or time), boom microphones are very useful for recording sounds in hard-to-reach locations. For example, they are great for recording the sounds of a faraway bees nest. Often hard to reach and definitely threatening, a 15-foot boom could place your microphone right in the thick of the activity.

Recording Public Events

The recording of lectures and speeches also benefit from lapel microphones or microphones connected to a headpiece. This application becomes leagues easier when one uses a lapel microphone connected to a radio transmitter. Wires tracing all of the way back to the camera or to a portable recorder are not ideal when people are moving about. With such a setup, you locate your camera on the other side of the auditorium, plug the receiver sound transmission into your camera or portable recording device, and record great sound.

When might you be able to avoid locating a microphone on the presenter? Large conferences, weddings, or concerts often broadcast sound through house speaker systems. Event planners provide the lecturer or maid of honor with a microphone that pumps their words through a soundboard and then the event hall's sound system. Sure, you could try to record the loud, echoing sound provided by these speakers, but we videographers have a much more elegant solution: plug a portable sound recorder into a "line-out" from the event's soundboard. You are now hijacking the person's microphone, originally used to reach the audience. This way the event planners are doing your dirty work – recording the sound up close and with fidelity. Later, you can sync the sound from your portable sound recorder to video clips that show the individual speaking (Image 13.5).

Note Lectures and speakers often talk with louder voices than they would use in a normal conversation. Because of this, you might have to turn the volume sensitivity of your microphone or camera down. Fidelity is lost when levels are too loud, which is termed *hot*.

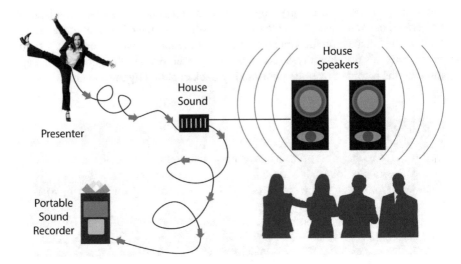

Image 13.5 When filming live events, where sound from a speaker or group of musicians is broadcast through "house sound,"use "audio out" connections (off of their mixer) to record sound. Usually mixers are delivering better acoustics than you could capture with your camera on portable sound recorder picking up ambient sounds within the room

Successfully Filming in Variable Volume Environments

Sometimes the sounds that you hope to record come in at a couple of levels of volume. One sound is loud and the other is quiet. This presents a dilemma. What volume settings should you use for, say, a shotgun microphone? If you adjust your audio settings on your camera to capture the quiet sounds, then the loud sounds come in too hot (i.e., loud sounds are recorded as garble). Oppositely, setting your sensitivity to the loud sounds means that you will miss the quiet ones. One solution is to use two microphones – one to capture quiet sounds and another for loud. Is this possible when you hope to record sound through your video camera? Some higher-end, prosumer cameras give you the option to record sound at different levels on two channels. What are channels? Most modern recorders capture on both a left and right (L and R) side. Each side is called a channel. When both sides are used to record a sound, the resulting sound track is termed *stereo*.

Some higher-end cameras give you the power to define what is recorded on each channel and come equipped with two audio inputs (usually XLR connectors). This differs from cameras with single microphone inputs as you'll likely find on low- to midrange gear. With two XLR audio inputs, you can use two microphones to capture very different sounds. This setup makes it very easy to drop the sound levels on one channel, such that the recorded sound has tolerances set to correctly capture louder sounds. Importantly, the second channel is then dedicated to recording more quiet sounds. One channel for the quiet sounds. One channel for loud. Setting your camera this way will give you options to choose (later, while editing) the best

Image 13.6 XLR audio adapters

channel recording for use in your final film. The loud one or the quiet one? Maybe a combination of both?

Since we are discussing repurposing channels that are typically used to create stereo sound, this new application may result in slightly imbalanced sound. By imbalanced I mean that to listeners, most of the sound may come dominantly from one side or the other. This can be somewhat corrected in your editing studio. There, you can manipulate your channel to sound centered, rather than dominating one side. This process is often referred to as "pan to center."

Portable sound recorders and mixers give you more options. Let's say that your camera can only record sounds at one level. Toting along a handy portable sound recorder can fill in where cameras fail to capture the sound levels you want. Portable sound recorders come equipped with their own microphone; however it is also useful to purchase one that has one or more 1/8th or XLR input jacks. The recorders with two inputs are usually closer to a few hundred dollars. Through these jacks you can plug in auxiliary microphones that will suite your needs of simultaneously capturing sounds at different levels (Image 13.6).

For as little as $100 you can purchase a very compact sound mixer. This is not to be confused with a portable sound recorder. Sound mixers take in sound from multiple sources and funneling them down to one track. This final track can be recorded on your camera or portable sound recorder (Image 13.7). Along the way, volume toggles let you correct the sound levels that each track provides for the final, outgoing track. This means that you are not recording sound onto separate channels, to be adjusted once back in your studio. Differently, it is your responsibility to monitor the sound levels that your system is recording in real time. If you don't like the sound of one microphone, turn it down while increasing the volume on another. In real time, you can then balance sound coming from different microphones.

Image 13.7 Mixers combine tracks arriving from numerous sources. With a mixer you can adjust the volume levels on each track. The resulting synthesis is recorded on one track

Sound mixers are very adept at balancing sound originating from different sources, each being captured by their own microphone. Concerts are great examples. You can dedicate a microphone to each instrument. By monitoring the sound that is leaving the mixer, you can turn up or down the volume levels from each source, perfecting your recording. Working effectively with a sound mixer is an art that takes practice and constant attention. If you know that you are presented with a fairly simple scenario, it is possible for you to run a camera and mixer alone; however, splitting the duties with another usually locks in better recordings.

Note If you plan on using two different microphones without a mixer, and instead record each microphone onto its own track, it's important to make use of a clapper. Right after pressing record on both microphones, use a clapper (or just simply clap your hands) to make a loud, impulsive sound that is later used to sync the audio tracks in postproduction. The syncing process is discussed in Chap. 16.

Recording Narrations

Narrations often serve as the backbone for sharing verbal information to a viewer. Typically, narrated videos include a commanding, all knowing, or third-person voice describing an event or an object, supported by descriptive video content.

Think about filming assembly directions for furniture or equipment. One could speed up a time-lapse of building a wheelbarrow purchased from a local hardware store. In the background, a voice describes the different steps, roadblocks, and uses of the various pieces being attached. Another example is a voice describing the merits of a certain public policy, while at the same, images portray past examples of where this policy excelled. Even DIY (do it yourself) instructional videos on the Internet use narrations. In these cases, the wielder of a smartphone camera can film an object while narrating offscreen from behind the phone.

The following are steps for improving the recording of narrations.

One: Write A Script

Precise words and sequencing lend impact to films driven by narrations. If this is your plan, make sure that the words that you say convey exactly what you want. Fortunately, a script gives you the choice to control your phrasing. Scriptwriting takes time and attention to storytelling. The point of a script is to carry a viewer logically and methodically through the elements of your story, and ultimately your film. Take time to organize the information that you would like to present so that the audience can follow it. I like to (1) write scripts with the needs of my intended audience in mind, (2) reread my work to verify that it flows like spoken tongue, and (3) test-drive the wording on friends or collaborators.

Two: Inject Tone

Think about the stories that were read to you as a little child. I think back to the voice, pacing, and inflection of my mother. She worked hard to keep me interested in stories by matching the qualities of her voice with the tenor of the story. Try to do the same thing with your narrations.

I heard from a fellow filmmaker that their most successful narrations, even for technical information, were often read as though it is being explained to an eighth grader. I took this comment to mean that a narrator should use more emphasis and emotion than one would consider appropriate compared to talking to somebody face to face. The energy level presented should not be pinned at high excitement, and oppositely it cannot be monotone. Think about how you can use inflection in your voice to reinforce your take-home messages. If you're explaining something sad, let the tone of your voice drop at the end of a sentence. If there's an important message, end on a sharp note. If you are imitating something that a person said, perhaps try to characterize their voice.

Three: Record with a Quality Microphone

Since sound is going to carry the bulk of your story during narrations, record the highest-quality sound possible. This builds your story on a robust foundation. A beautifully written script recorded with a low-quality microphone or with distracting background noise can result in selling yourself short – right out of the gates.

Therefore, the first step that you should always take is to record in a room that has minimal background sounds and echo. The next step is to think about what microphone is best. Just like cameras, you can spend a lot of money on microphones. However, since most of our discussions revolve around cost- and time-saving approaches, let's first discuss your simplest choices for microphones. For example, it is not unheard of for somebody to narrate into a lapel microphone. Just the same, in a pinch, people have turned to the built-in microphone that comes with DSLR cameras. As with all recordings, locate the narrator's mouth close to the microphone, being careful to keep their breath from hitting the microphone and creating disturbances in your recording. These options are suboptimal, however have been known to work.

Personally, I prefer to narrate into auxiliary microphones designed for podcasts. The sole purpose of these devices is to capture a great voice sound. With a little research online, you can uncover great quality podcast microphones for only a couple hundred dollars. Indeed, these microphones normally come with windscreens, but I like to supplement this protection with a pop guard. A pop guard is a small disk spanned by a thin mesh material. Normally a narrator will place the pop guard between their mouth and the microphone, usually separated by only an inch or two on either side. Should you, or any other narrator, breathe out too hard near the microphone, the material will stop the air from hitting the microphone and degrade the recording (Image 13.8).

Four: Record a Couple of Iterations

Although you might think that your first attempt at a narration was a smashing success, record the same reading for a second or third time. Experiment with different tones and emotions. Typically, once back in the editing studio, I look past my first recording because I, or any other narrator, are generally more confident in the script during a second or third attempt.

Image 13.8 Pairing a pop
guard and a podcasting
microphone is great for
recording narrations

Wind Interference

Wind interference can ruin almost any sound recording. Irrespective of the source
(atmospheric wind, a breath, or the microphone passing quickly through still air),
air movement over the microphone creates low-frequency disturbances that may
overlap with or completely occlude the sounds that you are trying to record.

How does interference form? Most microphones function by transferring sound
waves into an electric signal. One of the more common ways to do this is to suspend
a coil of wire within a magnetic field. Sound waves cause the coil to vibrate and thus
create an electric field. Interference forms when pulses of air masquerade as sound
waves and actuate the moving parts of your microphone. The qualities of the interference have much to do with the direction of the wind in relation to the
microphone.

There are several steps that you can take to reduce, if not remove, these deleterious side effects from a filming session.

Moving the Microphone Out of the Line of Fire

The first line of protection is to reduce the unwanted wind from reaching these moving parts. In many circumstances, a microphone can be hidden or buffeted from wind. For example, if you are filming a landscape and hoping to capture the ambient sound, consider hiding the camera, equipped with a microphone, behind a wall, body, or car. In some cases, you can cup your hand over smaller microphones creating a quiet air pocket. Similarly, if you are conducting an interview with a lapel microphone, angle your subject's body such that their back is to the wind.

Supplemental Microphone Guards

In other circumstances, it is very difficult to hide a camera from turbulent air. So hide the moving parts of your microphone from wind by simply moving your subject closer to the device. Fortunately, most types of microphones come with some level of protection against wind such that decreasing the distance between your subject and microphone will help reduce wind noise somewhat. In high wind situations, though, these protection methods might not be sufficient. Because of this, buy supplemental protection. Most microphone makers or third-party companies offer more stout windscreens then those that come stock with your camera or microphone. These are often made of foam or long strands of synthetic hair. Even more effective are Zeppelins. These are balloon-like orbs that surround shotgun microphones. In most cases, increased protection from wind comes at the cost of more prominent or unwieldy microphones. Lapel microphones appear larger and more obvious when worn with a wind guard (Image 13.9). Additionally, shotgun microphones can more frequently intrude into corners of shots when positioned on the hot shoe of a camera.

Creative Solutions

Some of the best solutions come by working creatively with what you have. Perhaps you are not interested in having your lapel microphone seen. Clothing can provide another form of cover from wind. For example, lapel microphones can be threaded up the back of an interviewee's shirt, over the top of their head, and poke out just beneath the bill of a baseball cap. When positioned correctly no one will ever see the microphone. Other options are to tape a microphone to the inside surface of their shirt, so that the microphone is positioned between the material and the skin. When done correctly, these methods work like magic. However, they come with complexities. For example, when a microphone is fixed to the inside of a shirt, rubbing sounds may manifest on your soundtrack. As always, practice your methods before your first day of filming.

Image 13.9 Additional windscreen protection can be purchased fit over microphones to decrease wind's deleterious effects. Here is a small wind cap, ideal for lavalier microphones

Note Always make sure that your interviewee is comfortable with any method of microphone placement.

Electronic Filters

Some cameras and microphones give you the option to digitally minimize the amount of low-frequency sound being recorded. With the flick of the switch the rumbling can be omitted, however, along with other low-frequency sounds. While this option can work wonders when on the run, the omission of low-frequency sound can leave a track sounding hollow and lacking dimension.

Concluding Thoughts on Recording Sound

Sound is an immensely important component of a film. No matter where the sound is coming from, the information that it is providing viewers either adds or subtracts from their experience. Viewers may never know why they liked your film so much, because great sound is rarely celebrated by audiences outside of the field of film making. Taking command of the qualities of sound that you are recording is a large step towards building your films on a hearty foundation. For this, you may become well recognized as a gifted videographer because of your attention to sound, yet people may never be able to pinpoint why.

Chapter 14
Video Editing: Postproduction

You Are in Postproduction Now!

The next few chapters are devoted to video editing. We will discuss the ways that most video editing programs (VEPs) function and then jump into how to cut, splice, and transition pieces of media. Sound can never be overemphasized, and so we'll delve into the art of audio edits in a chapter all to itself. Note: even though visual and audio edits will be discussed separately in this book, edits within your VEP can be conducted in conjunction with each other.

Organizational Scheme

I begin any large film editing undertaking by reviewing the mission of my film. This usually starts with rereading my storyboard. What next? I organize all of my clips into folders such that they can later be easily selected and intuitively inserted into the video editing timeline. A *timeline* is where in your editing program you compile and sequence all media together, such that the end result is a complete film. Even if I start the process of editing my film with one, long clip, like an interview, I often seek additional footage down the road. Why? To add cohesion, excitement, examples, or information for impact. More typically, a film that I am producing is the compilation of tens, if not hundreds (maybe thousands), of clips, images, songs, and figures. A necessary step is honing a file management strategy that makes sense.

Upfront careful planning about which files go in which folders invariably saves time and frustration as disorganized media will leave me chasing down jumbled media files. With a few films under your belt, you'll gain an appreciation for which organizational strategies work best for you…and conducting the file management early on in editing will seem second nature.

How might I organize footage? Let's say that filming produces footage of several different landscapes during the 12 months of the year. I would likely create folders

© Springer International Publishing AG, part of Springer Nature 2018
R. Vachon, *Science Videos*, https://doi.org/10.1007/978-3-319-69512-9_14

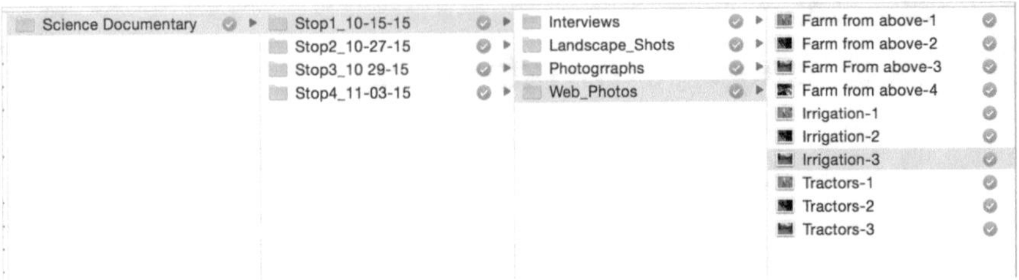

Image 14.1 An effective first step in editing a film is to create a well-organized and structured file management scheme. This way, when you search for the files that will constitute your film, you can readily find what you are looking for

defining the location and the time of the shots, like "Barbara's Farm-December." All files that pertain to my days of filming at Barbara's farm, during the month of December, are dragged into that folder. I would then create folders for all other months of filming. I may open additional subdirectories within such folders that hold relevant photos, articles, interviews, or B-roll (Image 14.1).

Some folks go even further to name every video file in a way that describes what they embody. "George_Fertilizers-1" might be the first file where George describes the value of fertilizers. Personally, I don't like to remove the original file name of the film clip. This is especially true if I back up all of my original media elsewhere. This way I can refind files should newer versions be corrupted or if I would like to use the files for another project. That said, I can find use in ranking similarly filmed clips by how useful I think they are. This is a step that most helps when I am working with a large number of clips that have been sitting unused for long periods of time. Let's say that during a filming session I recorded several takes of a given scene. Later, I will add descriptive notations identifying quality, application, or appropriateness of each clip. This way, when I am editing, I don't have to sort through all of the takes again – amounting to wasted time. If the original file name for a clip was IMG5460. I would call the third best take for scene 5, "5-3_IMG5460."

Above and beyond media files are editing files. VEPs create editing files which contain the coding that connects together all media files, transitions, volume/color changes, and titles. This brings up an important point about how VEPs function. VEPs do not contain original media. Instead, original media stays in their original folders and are referenced by the VEP. Perhaps think of editing files as the glue to hold the trimmed media of a film together. How the files relate to each other is defined by the editing files. Once the editing file, with all associated media, is exported, you will have your cohesive movie.

At the outset of editing, verify that the VEP editing files, associated with your project, are in an easy-to-find location and stay there. If you move the location of the VEP editing file or your media halfway through the editing process, there is a good chance that you will have to relink how the VEP finds the media. Depending upon the VEP and the amount of footage, this can be very time consuming.

Video Files Take Up Space

I would say that the greatest hurdle to editing films is the data volume required for storage and processing. It can be very large, if not huge! If you are working on several films, or one large one, the volume of data can fill up entire hard drives on some high-end laptops and bog down RAM. To overcome these challenges, pick your data storage workflow carefully, particularly as you get involved in filmmaking projects of mounting complexity.

File management strategies are rapidly changing with increasing memory for hard drives and fast-paced internet speeds. Some people save all media files on the Internet, such as with Apple's iCloud. Such solutions for storage are great for working with teams, where footage can be accessed by all. Unfortunately, with large video projects these solutions can come at the cost of maintaining web data storage accounts, and you must always ensure you have the proper bandwidth to access the files. Alternatively, you can consider storing media files on a separate hard drive. When working in remote settings, like developing countries, working off an external hard drive works well. As a note, if you choose to store your footage off of your computer, either on a separate hard drive or on the web, data transfer rates are often slower compared to those on a solid-state memory hard drive that comes standard with many new computers.

Finding the solution that will work for you can take some research. I would suggest talking to other filmmakers who are making films similar to yours and learn how they manage their work.

Warning Video files are complex and large, and editing programs are intricate, which means that they are prone to crashing. Not only should you save your VEP editing files often, but backup all of your media assets regularly. Redundancy will save a project that suffers the fatal losses that occur from time to time.

Unfortunately, backing up large files is time and computer processor intensive, so dedicate enough time to have your computer connected to a drive that will adequately copy and handle your files. Note: leave backing up files to periods when you will not be burdening your computer with processor intensive activities, like video editing or exporting. Nighttime, coffee breaks, answering emails, or lunch suit these needs nicely.

First Pass: Leaning on Your Storyboard to Build Your Videos

Now it is time to delve into stitching your media together into a story. As we progress through this discussion, we will call your first version of your film your *first pass*. Building your first pass is often a very intense time for beginning film editors (heck, even for professionals). There is so much work to do, especially with large films! Plus, new learners are coming up to speed on how to use their VEP and might not be aware of all its capabilities. Much like learning how to use your camera, start slowly, learning the basics and being careful to avoid jumping into rabbit holes of complexity.

How do you start with editing films for impact and efficiency? I immediately turn to my storyboard as a guide. If completed correctly, your storyboard will provide great directions for how you can drag specific media to certain points on your VEP's timeline. Your storyboard might not tell you exactly which shots fit where, but it should give you, at the very least, a good feeling of what you are looking for. With time and experience, you will learn just how valuable your storyboard is for staging your film-building efforts.

As part of the jitters that come with starting a film, some introductory videographers begin editing concerned that their efforts will not result in the impact or effect that they desired. If this is you, I highly suggest that you take a step back from the end goal of your film and begin editing systematically. The vision of your story is locked in your earlier efforts on your storyboard. Maybe it is helpful to think about the process like the writing of publications or progress reports. The critique of a rough draft is best done when you have complete products or chapters.

Here is another suggestion. Take on a mind-set that you are a contractor working for someone who has specific hopes for a film. They need you to interpret a storyboard and manufacture a first draft – connect the pieces, nothing more. Your job is to use your storyboard as a straightforward recipe, a great starting place for dragging in the media that you think best represents what you earlier envisioned.

Keep modest expectations for your first pass. This is your first rough draft and it will likely be just that. Editors usually cite that they become very productive judges of their work after the first pass. Thoughtful next steps to produce greater impact emerge. Know that fine edits, transitions, and the inclusion of music, later in the game, will give your film grip!

Taming the Need to Be Creative

When starting to edit a new project, it is easy to fall into the creativity trap. What if you have started editing from your storyboard, but inspiration to be creative leads to thoughts that your storyboard is restrictive? Perhaps you have recently seen other people's videos that motivated you or new ideas blossomed as you produced your media. Maybe this has left you believing that you can improve upon your original idea and storyboard. Now you are tempted to jump off your storyboard headlong into a very creative voyage of weaving your story together from the notions in your head.

No matter the reason, thoughts rattling in your head are very valuable and deserve consideration for inclusion in the final video. However, an immense amount of time can be lost by trying to configure clips in your VEP based upon whims in your head. Why? Because the ideas do not often have the greater mission of your film at heart. Story building takes time and nurturing, and spontaneity can lack this backbone. Many storytellers agree that creativity must be heard in balance with the ideas that you developed earlier in your storyboard. Potent stories about technical or scientific content are a combined effort between both sides, linear and nonlinear, of the brain. Experience will guide you in how to strike a balance that works for you. However, it is helpful to start editing in more time-honored ways. Here are a couple of ideas.

Approach #1

First, make an agreement with yourself that you can follow your inclinations towards being spontaneous, unstructured, and creative for a dedicated afternoon. Weave a new story with your media. Try developing fascinating effects that will, say, carry a viewer between scenes in a catchy way. However, in concert you must also allot the same amount of time to follow your storyboard in a direct manner. This dual approach has saved me from the rabbit hole of excessive creativity which may ultimately prove to be distressingly diffuse and unstructured. The dedicated time puts bookends on the spontaneity.

Further down the road, it will be necessary to decide whether your new ideas produced fruit. Perhaps your new ideas were silly. Maybe they were leagues better than your original storyboard. Usually a good sleep helps you figure out whether you should continue on the exploration in new ideas. What do I usually find? The creative period of this exercise opens my mind to new ways of building discrete segments of my preestablished story. If I am lucky, I can cut and paste elements from the experimental effort into the previously conceived film.

Many editors have a notebook beside their keyboard (or a *Word* document opened in another window), dedicated to project notes. These are perfect places for jotting down thoughts as we review, ponder, and organize our files. As ideas or tangents crop up, simply write them down. Later, as we construct the film in the VEP, we can return to them and fold them into the storyline. Personally, I find it productive to go through the act of taking notes because it slows down my thinking and formalizes the procedure of reviewing and processing my media assets.

Approach #2

A second option for satiating the need to be creative, while giving your storyboard the chance to be the tool that it was initially designed to be, revolves around revisiting your storyboard with heightened insight. Make a duplicate copy of your original storyboard and then read it over. How does it sound? It can be exciting to go back and see just how ingenious you were and how you might be able to conjure images or words more powerfully than you originally thought. Maybe what you first designed shines through as the obvious way forward. More likely, you will see room for improvement. Take notes on where you think that it could be improved, particularly through the lens of your new insights. Once you are charged with a refreshed understanding of the mission, hammer away at the storyboard, tightening, improving, bringing it to new life! Chances are your new creation will be some fraction of the original plan. ⅔? ¾? It all depends upon how revolutionary your new thoughts were.

If you are going to attempt the second option, I suggest focusing your efforts down to completing revisions to isolated segments or acts. This way your efforts to inject creativity or a more potent workflow are not left hanging as unfinished pieces. It can be very disheartening to go into an effort with great intentions, but only to come away with a lost result.

Introduction to Editing

VEPs are the canvas for your film – the place to paint an elegant story that conveys the feeling, knowledge, or perspective that you have in mind. The more versed you are with your VEP, the more quickly and potently you can build films. Experience will give you a deeper appreciation for the tools that you can use to bolster your storytelling.

Entering into this conversation, take a moment and return to the part of this book that goes over purchasing editing software. Maybe over the course of reading this book and discussing your film with other folks, you are leaning towards a more expensive VEP than you originally thought. Conversely, maybe filmmaking sounds like a huge time sink and you want to start with a simpler VEP.

If you need a reminder – inexpensive programs will typically limit how refined you can get with transitions and filters (for color and sound). Inexpensive VEPs do come equipped with some simple tools, though, that may give certain edits a classy feel and render others cheesy. Inexpensive editing programs are the canvas for incredible movies every day, but are generally great for someone who only wants to make basic videos. Alternatively, more expensive programs have steep learning curves, but once you become proficient, you can quickly assemble films of high value. If you have hopes to produce several films, more expensive VEPs give you an enormous number of options for creating a specific feel to your film or fixing less-than-ideal video and sound bites. They also give you the ability to use multiple forms of media (e.g., still images or graphs), as well as multiple file formats (e.g., .AVI or .MOV).

Note In keeping with earlier discussions about file formats, verify that your editing program can be paired to the video format of your camera. For example, if your camera exports in .MOV, make sure your editing program works well with these file types (most do with .MOV). Changing files to different formats, so that a VEP can wield the files, typically means that the quality of your video clips is diminished. Problems with file compatibility are not usually a huge deal these days, but used to present big problems to progress. Perhaps this note comes from scars of my own past; however, I suggest double-checking to be sure.

Navigating Your VEP

There is no substitute for diving head first into a VEP and struggling to figure out what tools work best to convey your vision. As you begin to mess around in your VEP, I would like to offer pathways to avoid enormous time sinks or lost efforts as you gain experience with the ways of the road. I would also love for you to learn how different edits illicit given tones within a film. I will share a number of approaches and techniques, in broad-brush strokes, to pilot you into more stream-lined assembly of functional movies. Fair warning: even with direction, implementing lessons in video editing can easily spin into a process that takes five times more time, emotional energy, and resources than expected.

Unfortunately, the host of VEPs that you have to choose from have different user interfaces. Most programs conduct very similar procedures, but the icons, placement of buttons, and adjustment knobs differ enough to make it next to impossible to have a detailed discussion about how to conduct a given step for all VEPs. Because of this, you will read *often*, *a lot of*, and *many* a number of times in this chapter. If you would like to learn specific methods to editing within the VEP that you choose, consult the user's manual for the VEP. There are countless VEP-specific, online tutorials showing you how to perform precise edits. I would suggest that you read the following chapter and then take a lunch break, to search just how rich the online resources are for your VEP.

VEP Architecture

Here we will give a very general account on how to navigate most VEPs. I will also go over the fundamental video editing tools that equip most VEPs. That way, you can come away with a clearer understanding of the power, pitfalls, and potential of most VEPs.

Let's break down the architecture of most programs. I'll try to answer – where can you organize the raw media files that you wish to use in your film? Where are your media assets sliced and sequenced to tell your story? Where can you find tools that make specific changes to your video and audio clips? Where can you preview what your movie will look like? Most user interfaces within editing programs include a few standard regions or zones where each of these procedures happen. Much of these customary designs come from the long practice of movie editing that first began with film, not digital files. For this reason, once you learn how to edit with one VEP, it is often fairly easy to transition over to another program.

Project Window

Programs typically have a space for you to catalog the files that you hope to use in your film. This is often referred to as your *project window*. These are places where you can drag in what files you are using to tell your story. Again, you are not changing the location of these files to reside in your VEP. This is simply one way for you to tell your VEP that you will be or are using these files. Whether a photo, soundtrack, title, or video clip, each type of media likely will take on a look that makes it identifiable. Perhaps they are as simple as a postage-stamp image from an instance in a film clip along with the file name, or sound waves representing an audio track. If you click on the icon within your *project window*, you can review the entire piece, not just the shorter segment that you might include in your edited film. Without a

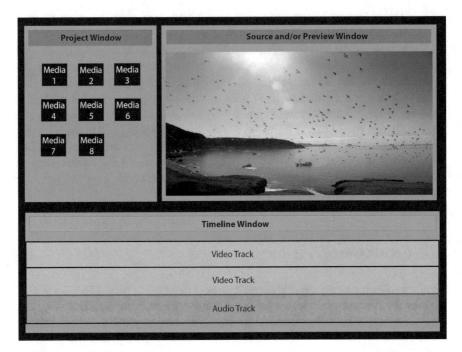

Image 14.2 The architecture of a typical, yet simplified, VEP screen. Note, it is often the case that a source window is interchangeable with a preview window

p*roject window*, you are forced to go to folders within your database every time that you want to access files (Image 14.2).

Source Window

You may be able to review the clips that you would like to use in a *source window*. For this you can click on your assets in your project window and the clip will become active in a separate window. More than review, you can define how much of the clip that you would like to include on your timeline. For example, you can define that only 5 s, of a 16 s raw clip, be used for your story.

Timeline Windows

A *timeline window* is likely the most recognized space of a VEP. Your timeline is where you connect all of your trimmed media assets chronologically. On your time-line, media is linked, one after the other, after the other. If your film will be 10 min in length, this is how long your timeline will be.

The *timeline* may be divided into two zones – a region of tracks dedicated to visible media (like photos or video clips), situated above a dividing line. Below this line is a region dedicated to audio tracks. Typically, the zones assigned to video and audio assets have space for several tracks. Multiple tracks mean that media files can be stacked upon each other, such that several can be played simultaneously. Most importantly, as you drag or insert visual clips onto your timeline, the topmost video or image files will overlay the other tracks. For example, if you have an arrow image stacked upon an image of a mountain range, in your preview window the arrow can point to the top of the mountain of interest. Changing the transparency of an overlapping video clip can create another very cool effect for your film, whereby you partially see through your top layer to see one beneath. This option is also very useful for testing which of two clips strengthens the appearance of your film. Turn the top layer off to see your second option beneath – a simple procedure for experimenting with appearance and function.

Soundtracks also stack on timelines. Oftentimes, a soundtrack is linked to a video clip. This is how clips are formatted in your camera. When you insert a video clip into the region dedicated to videos, a linked audio track also inserts into the region dedicated to soundtracks. The higher up the video clip stands in your video tracks, the lower the associated audio track locates in the list of soundtracks. While soundtracks also overlap in a timeline, they behave differently from video clips. How? One clip is not given priority over the others. As a result, when soundtracks are stacked, they all play together. If a song is lower on the timeline than the sound of feet running on gravel, both sounds will play at the volume that you define. The result can be complementary or muddling. Fortunately, VEPs give you the control to turn on or off, or change the volume of soundtracks to optimize this interactivity.

Preview Window

It is valuable to preview your timeline. If you locate your cursor within your timeline, and press the spacebar, a vertical bar will progress through all media assets compiled on the timeline. All of your edits should play in a dedicated window as they will appear and sound in your final video. Where you view the preview is called your *preview window*. The speed of playing generally happens at the rate at which your clips were recorded (unless you deliberately speed them up or slow them down). This is also the speed that a viewer will experience your final video.

Note This is only true for clips recorded at the same frame rate as your final film. However, if you mix and match clips of variable frame rates, some clips may play on your timeline at different speeds than others. For example, a clip recorded at 24 frames per second may play more quickly than other clips that were recorded at 30 frames per second.

Since you might be mixing and matching file types, formats, and filters, the resolution of these playbacks is often purposefully reduced in the *preview window*. This means that your preview looks grainier than your final video. This is a benefit since a lot of media compiled, edited, and played at once can bog down your computer to the point that the playback may appear stilted or inaccurate. Appropriately, more refined editing programs give you the option to adjust the resolution so that you can preview your video in the most true-to-the-final-video flow that your computer can muster. Indeed, more powerful computers handle complex previews more easily than slower ones.

The amount of time visible on your timeline (the space across your screen) can be expanded or compressed to capture longer or shorter amounts of time. This allows you to adjust how many clips, from within your story, are visible on your screen. This option allows you to make your timeline most functional for your uses. For example, seeing all of the clips in a 20-min film makes it easy to move large bodies of clips from the beginning of your film to the end. You can navigate large portions of your video rapidly. Once you click on a region of your timeline that you want to explore in greater detail, you can then zoom in for more detailed viewing and editing. In seconds, you can go from a long, 20-min video filling your screen, to only a 5-s microsegment. With a shorter segment filling your window, you can scroll to the left or right, moving towards the beginning or end of your film, respectively. Seeing these portions of your film in more detail facilitates refined edits, such as shaving off fractions of a second at the end of an interview clip (Image 14.3).

Track Management

It is useful to create a track management scheme within your VEP. This is an organization approach so that you know into which tracks on your timeline to place and find certain forms of media. Progress becomes more efficient when you don't

Cool Trick
I last mentioned how you can zoom in for refined edits of an interview clip. This is a great chance to descend into a technique that excels at these scales. What if you are editing an interview and wish to end an interviewee's words at a certain point, but the interviewee went on to say much more? It is useful to cut off the clip before their next breath (which often follows at the end of a sentence that is about to lead into another). Leaving such inhales at the end of a clip are dead giveaways that the interviewee had more to say. When audiences hear an inhale, yet don't hear a follow-up comment, they may feel like they are left hanging. You, the editor, can cut off that breath with precision. Subduing or lowering the volume at the right time of such clips is a careful edit that is more easily undertaken when you zoom way in on your video/sound clips.

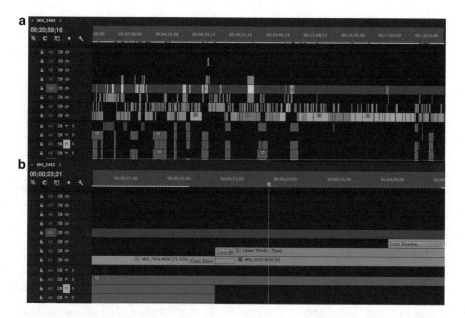

Image 14.3 Timelines in VEPs can be expanded (Image **a**) or compressed (Image **b**) for increased functionality (using *Adobe Premiere Pro*)

simply drag media onto any random track in your timeline. This causes files to become jumbled. The time savings of devising a track management scheme usually manifest later on in the editing process.

What might such a scheme look like? When editing a documentary, I often dedicate my first video track to interview clips, my second video track to field-based B-roll, my third track for photos from external media sources, and my fourth track for titles and shapes. The most important message to take home from this conversation is you will benefit from using a scheme that is best devised through your own personal experiences (Image 14.4).

Start to Build on Your Timeline, But Don't Bite Off Too Much

A healthy place to begin the editing process is with realistic expectations of the time and effort needed to completely edit a film. These estimates are based upon the quality of your storyboard, ownership of skills, and deep familiarity with the media files. When films are quite large, I like to then micro-unit my efforts into smaller goals – editing shorter acts that fit into the larger scope of my film. If I hope to produce a 1-h video, yet an 8 min act is dedicated to one succinct part of the story, then I focus all of my creative and structured attention on editing that one segment before moving to another.

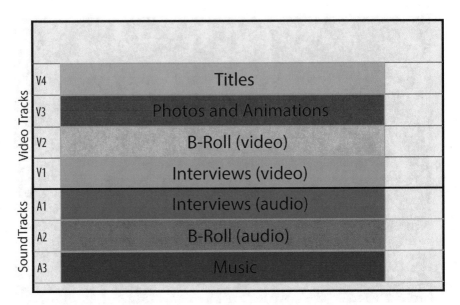

Image 14.4 Large video editing undertakings are best tackled when you devise and stick to a track management scheme. Within your VEP define what type of media belongs on each track. The example shown here is the track management scheme that I use for editing documentaries

I think that sticking to editing together singular acts is useful because some film-makers can slip into wanting to assemble a large and complex story straight out of the gate. In an effort to piece it all together in one large push, they can do quite the opposite. Without appreciable tasks being accomplished, the greater purpose of the film may be lost, and apathy sets in. Defining and striving for lesser goals (on the path to your complete film) can lend satisfaction through measurable progress.

Inserting Files into Your Timeline

There are two things to think about when learning to insert files into your timeline. One is how you approach assembling the clips that best build your story (defined by your storyboard). The other is the method for embedding these clips into the appropriate place on your timeline.

I will talk about several editing tools that are useful once your story is assembled shortly, but right now I'll focus the discussion on two ways for inserting files from your folders into your timeline. For both methods, drag and drop all of the media files that you think will be useful for your film from hard drive folders into your project window.

Next, bring these assets from your project window into your timeline. For some programs, you can define precisely where you would like each clip to begin and end in the project window or in an associated source window. What I am saying is most raw clips are longer than what you need within your film, so you can select the portion of the clip that you want to use in your film before dragging it into your timeline. One can bring clips into the project by double-clicking on the media asset in the project window, thereby opening it for review. As you play your clip, you can then define where you enter and exit the clip with tabs. Usually there are keyboard shortcuts (such as the letters "I" for "in" and "O" for "out") that define the clip's beginning and end. This process of refining the length of a clip, before it is inserted into your timeline, is also called *marking*.

Now is a good time to mention *time coding*. Every digital clip that you drag into your project window has a duration. The duration of a clip is measured by its *time code*. Time codes are measured in hours, minutes, seconds, and frames. The last measure, frames, is a breakdown of how many frames were captured per second. A time code may appear like, 00:17:46:12, which connotes that you have reached the 17th minute, 46th second, and 12th frame of a clip. As you review a clip, the time code value will increase as you proceed. You can also find a time code associated with your timeline. In these cases, the time code is a measure of how far you have proceeded into your film. Let's say that you have five 5 s clips stacked end to end within your timeline. If you are editing at the 4th second of your last clip, your time code on your timeline will measure 00:00:24:00.

Now, you must choose how to insert clips into your video.

Option 1: Insert

The first option, *insert*, acts like a wedge, forcing a new clip between two preexisting clips that are now tip to tail on your timeline. With a clip chosen in your source window, place the cursor at the junction between two existing clips in your timeline, and press the *insert* button. In order to make space for the new clip, this tool pushes the existing media to the right of the cursor farther down the timeline. If there were no preexisting files to wedge between, this act will simply place the new clip on the timeline after the cursor (Image 14.5).

Option 2: Overwrite

Your second option, sometimes called *overwrite*, embeds the new clip into the timeline without pushing trailing clips further along the timeline. Your new clip will write over the clip(s) that follows your cursor. To execute overwrite, you are usually presented with an overwrite button (Image 14.6).

For both options 1 and 2, you must be very aware of where the new video and audio tracks will appear. This is particularly important for the second option (overwriting) – if the new clip is inserted on the same track as the preexisting clips, the older clips are overwritten. Likewise, if a new clip is inserted on a track above a preexisting clip, the new clip will mask the preexisting one in the final video. Awareness will help you decide how to assign the video track where the new clip

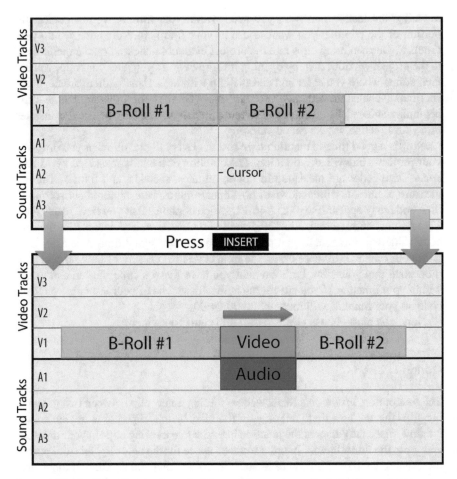

Image 14.5 The "Insert" option for embedding new footage into a film's timeline typically means that the new clip is placed following the cursor and any preexisting clips on the timeline are moved further down the timeline

will post. This way the overwriting may or may not necessarily have to eliminate preexisting footage.

What might this look like? Let's say that you want to paste a clip from a video interview under a couple of B-roll shots that show what is said in the interview. This way you will see the B-roll, but hear the interviewee's voice. What's more, you want to have the video of the interviewee under the B-roll, thus retaining (although ultimately not visible) the video component of the interview. Why? In case you later want to remove the B-roll and simply watch the subject speak. Let's make sure that you insert the interviewee video clip on a track *under* the B-roll. So, first relocate your B-roll onto the next video track up (so that it will be what you see), then position your cursor where you want your new track to begin. This will place the new

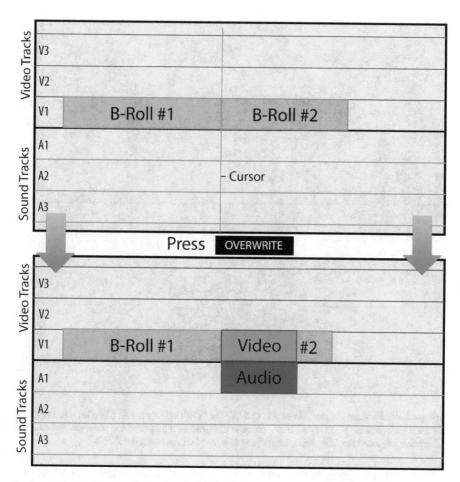

Image 14.6 The "Overwrite" option for embedding new footage into a film's timeline typically means that new clips are placed following the cursor, but any preexisting clips on the timeline are NOT moved further down the timeline. Preexisting clips will be overwritten unless steps are taken to make certain that the overwriting clips are placed onto uninhabited tracks

clip on the first video track. In the process, footage is not lost due to overwriting (Image 14.7).

Option 3: Click and Drag

Most VEPs afford you a third and more thuggish method for inserting clips into your timeline. Simply navigate to the folder where your video files are housed. Click and drag the appropriate files from the folders directly to your timeline. Not only will the full length of your video clip imbed on your timeline, but the clip will also appear in your project window. This step may sound like the most

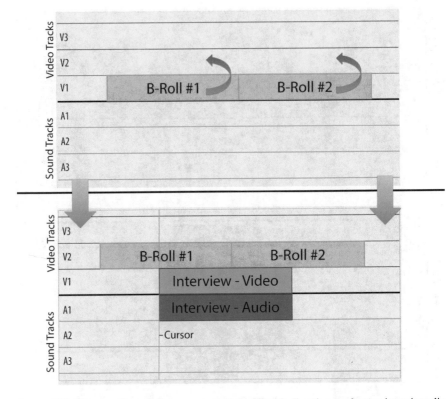

Image 14.7 Here we illustrate how you can use the "Overlap" option to place an interview clip under a sequence of B-roll shots. First, move your B-roll to a higher video track than where you will place the interview clip. Next, insert your interview clip beneath the B-roll

straightforward option, but it does not give you the chance to edit the length of you video clip before you drag it into place. If your clip is quite close to the desired length, this may not be a bad option, but it becomes very burdensome when you want to insert 2 s of speaking from a 10-min interview. In these situations, the click-and-drag option becomes an unwieldy choice for inserting files.

Inserting Files While Adhering to a Storyboard

Now that you know how to embed files onto your timeline, it is up to you to pick a logical approach for moving the media onto your timeline in a fashion that adheres to your storyboard. There are a couple basic methods to consider.

One approach is to methodically scour your folders, and as you come across a good clip, insert it onto your timeline where you think it fits. This approach is most effective when you use your storyboard as a checkoff list, whereby you literally check off the clip on your storyboard as you locate an appropriate segment. During this process, you might feel disconnected from the flow of your story, because the progressive movement through folders may not mirror a chronological progression through your storyboard. Believe in the process. Soon, when you preview your timeline, with all media in place, your story will emerge.

Should you lean towards using this method, at the end of the process return to your storyboard and read it in full. Verify that you have checked off all clips that you have embedded and note which ones are missing. This second reading of your storyboard should inform you on how to transition scenes, use music, control color, and elicit tone. This function hinges on how well you built your storyboard. Thorough construction includes notes or useful prompts to inform you on the steps that will result in the feel that you want.

The second option stages movie-building efforts from systematic movement through your storyboard. Step by step, identify and insert the clips chronologically through your story. This method ensures that you find all of the clips that you need as prescribed by your storyboard. Moving through a storyboard from beginning to end, while reading the notes attached to each segment, is an easy way to remind you how scenes are meant to appear, unfold, and feel. For me, this method leaves me feeling more in touch with the pacing of the story. I have seen this benefit shine through while reviewing my first-pass version of my film. However, this second approach can be very time consuming. Combing through countless folders to fill in your timeline chronologically is arduous, a problem tackled by well-structured file management systems (Image 14.8).

Regardless of which approach you use to insert your media, I find it most fulfilling to approximate the length of each clip first (according to what I need from my storyboard) and then embed it in the timeline (either with *insert* or *overwrite*). This concept needs a little clarifying. I trim the clip down to its most useful piece without focusing too much effort on exact edits and transitions. I save the precise tuning for later, when I am more certain that the media assets that I have chosen are best for the scene. Confidence in my clip decisions often comes only after I rough-in my best guesses and review the results a few times.

Editing on Your Timeline

Okay, you've got everything on your timeline, now let's talk about the fundamentals of editing! Clips on your timeline can be manipulated in a number of ways to best suit the needs of your film. Manipulations can include clip length, speed, color, volume, and more. Most VEPs give you a number of standard transition options that can be dragged into place. These are the tools that strengthen a viewer's experience, and thus connection to a film.

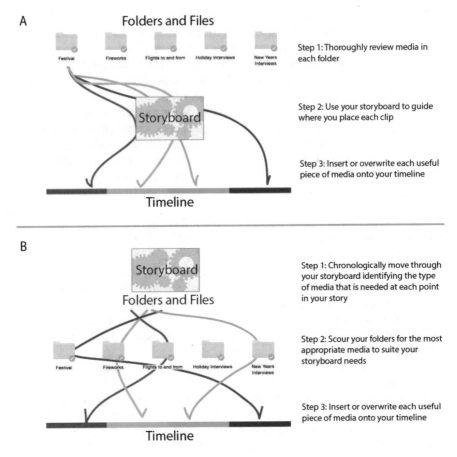

Image 14.8 There are two basic approaches for inserting clips into your film. (**a**) Thoroughly scour your folders for footage. As you find usable clips, place them onto the timeline in positions defined by your storyboard. (**b**) Work sequentially through your storyboard, identifying which clip comes next, and then scouring your folders for the best clips

Getting to Know the User Interface of Your VEP

Typically near and around your timeline (like on a toolbar near the top) are icons very similar to those that one can find on the toolbar in a *Microsoft Word* document. Each icon is a shortcut for some of the more common steps or edits. An example is an image of a disk that saves a file. Or you can change your cursor from a tool to select the clips of choice to a razor blade that will cut a clip at the point at which you place your cursor.

Learning the tools associated with the icons on your screen is one of the easiest ways to save time while editing films. For inexpensive VEPs, sometimes the links may be the only way that you can activate tools. Some VEPS also have keyboard shortcuts that serve the same function as the links. A more commonly known example is "Ctrl" + "S" which saves a file. More complicated VEPs allow you to customize how the icons appear and some of the default settings associated with each action. For instance, you can place a transition that fades from one scene to another. Usually transitions default to take 1 s, but you can change these settings to shorter or longer durations. Many programs contain more editing options than one can find among the icons. Read your user's manual to learn how rich and abundant your editing choices are.

Let's categorize some of the basic tools that you will likely find on your VEP.

Cursor

Just like a *Word* document, you can move a cursor around VEP workspaces with your mouse or trackpad. If your cursor is situated over part of your timeline, you can click and activate different pieces of media. Once selected you may change the length of the clip or move it about the timeline. If you click and drag the cursor over several media segments, more than one will become selected. With a host of clips selected, you can paste in attributes from another clip. An example might be if you wish to apply volume settings from one clip to any number of clips. Select the one clip that has the well-adjusted volume edits applied (Image 14.9). Copy the clip. Then select the several clips (that you want to have the same volume adjustments) and paste in the volume changes. The function for pasting in such changes is usually called something other than "paste" and more like "ripple" or "paste attributes."

Trimming Media

Usually VEPs provide two obvious ways to edit the length of each clip once it is inserted onto your timeline. First, select the clip in your timeline, and then place the cursor over one side of the file that you wish to trim or elongate. Click and hold, while dragging the edge of the clip in the direction of the desired change. This works for the leading or trailing edge of the clip (Image 14.10). Unfortunately, this step becomes more complicated when you wish to elongate a clip that has a following clip blocking this from happening. You either have to move one of the clips or use another editing tool that either overlaps the first clip over the part of the second file that is getting in the way or pushes the second (and all following) pieces of media further down the timeline. This procedure benefits from being zoomed in closely to the end of the clip that is being trimmed or elongated – giving the edit more precision.

Your second option for trimming a clip includes a tool called a *razor*. The razor chops a length of media where you select. The razor blade will cut at the portion of a clip where you want it to terminate. That way you remove the beginning and the end of the longer clip very rapidly. This is very useful when you have a clip that is, say, 15 s, but you only want the middle seven seconds.

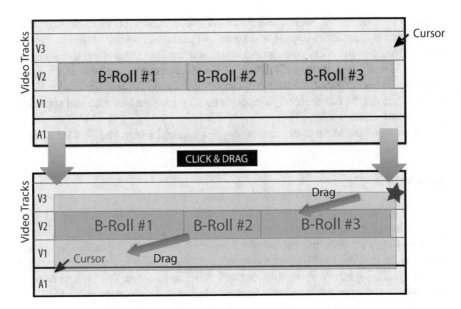

Image 14.9 An editor can select several clips at once, by clicking and dragging the cursor over the clips of interest. Once they are selected, you can paste edits, such as volume changes or color filters, onto all those that are selected

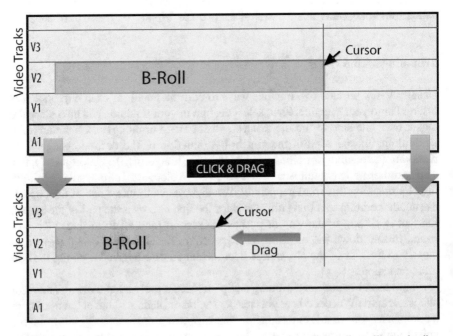

Image 14.10 One method for lengthening or shortening a clip is to click on the trailing or leading edge of the clip that you wish to alter, and with the button held, drag the cursor to where you want the edited clip to terminate or begin, respectively

Linking and Grouping

Sometimes it is easier to manage different pieces of media when they are connected to each other. For example, if you filmed an interview, most programs insert both the soundtrack and video tracks at the same time. If you click on, say, the video track, the soundtrack will also become selected. This way you can easily move both tracks around or cut them as one. When they are connected as described, they are *linked*. Oppositely, it can be useful to *unlink* clips. Consider how you could unlink video and audio tracks from an interview. With them unlinked, you can trim the beginning of the video track such that sound begins well before you start to see video footage (known as a J-cut). Once clipped, relink the two tracks and move the newly edited interview tracks around as one.

If you work on an entire body of clips and you think that they are edited together really nicely, connect all of the pieces together as a unit. The tool that usually accomplishes this act is *grouping*. When you group media, you can't accidentally move or delete pieces from the construct, thus destroying your earlier efforts to build a cohesive piece. Grouping may save you from losing precious minutes or hours of work.

Titles

One can construct clips where explanatory words can overlay a background, descriptive image, or video clip. This option is often used to build a title (Image 14.11). Similar to written documents, one is usually presented with a great deal of freedom to format the words (size, color, and font) within these titles. You can typically reposition the word to any location on the screen. This empowers you to label critical parts of a descriptive image. Titles also serve as subtitles or written translations of what a person is saying. For your reference, subtitling is typically centered on the lower portion of the screen. Simply said, the applications for titles are countless!

Titles can move or morph as a clip unfolds. Most people can identify with this function, where credits at the end of a film scroll up the screen. Likewise, you can have words pop up to reinforce concepts. Maybe you know of the old television show *Batman* (from the mid- to late 1960s). Words like "BLAM" would pop up on the screen to emphasize the force when someone was punched. You can do that! Even better, you can adjust the words to expand from minuscule to huge over the course of a second creating IMPACT!

Transitions

The simplest way to move from one video (or audio) clip to the next, in a timeline, is by trimming the first clip and then abutting it with the next. As the film unfolds, the first scene will instantaneously jump from one clip to the next. If the scenes are well composed and one clip feeds seamlessly into the next, instantaneous transitions

Image 14.11 VEPs give you a great deal of freedom to place text over background imagery. Such text is useful for labeling figures, presenting titles to films or rolling subtitles

Image 14.12 During a fade transition, one clip gradually disappears as another takes its place

are great. However, they can also be jarring. Fortunately for us, most editing programs come equipped with stock transitions. These can be dragged between two clips to help support the tone that you are trying to elicit with your video.

Cool Trick: Pretending to Be a Camera Flash
Turn a transition into a camera flashbulb! This is a perfect mechanism for transporting a viewer from one perspective to another, literally in a flash. This is a tool for integrating still photos into a film. Flash, flash, flash! You can share a rapid succession of illustrative images. This effect is easily fabricated with a "fade" or "dip" to white between scenes. Default settings have these fades occur over 1 s, but reducing the duration to closer to a quarter or half of a second gives a more realistic flashbulb feel. Personally, I find the transition more convincing when I shorten the start of the transition to white and elongate the exit to the next scene for a more realistic feel, still fitting into the 0.25–0.5 s interval. Now add the sound of a camera shutter and you have a very convincing flashbulb transition!

A great example of an effective transition is the fade, where the first clip gradually morphs into the second (Image 14.12). One suitable application for the fade is between breathtaking, dreamy vistas of national parks. Babbling brooks changing over to a time-lapse of clouds over soaring mountains. Another transition dips to a color between each clip. For scenes that tidy up one discussion and move to another, dipping to black can insinuate the conclusion of one idea and the beginning of another. Transitions can become far more complicated than this. A scene can flip, roll, or tumble to the next. With all of these choices, test how each affects a scene.

Note You can tailor your own transitions in some VEPs. I find changes to scale, location in the frame, and transparency are very powerful when coupled together.

Transitions can bring videos to life; however, the overuse of any of them drains the viewer. Just as important, complicated transitions are not necessarily better transitions. They can distract a viewer from content that a storyteller wants to be absorbed.

Video Effects

Video effects is a very broad category of tools for more subtle manipulation of your video clips. With more simple programs, you can click and drag prefabricated filters over your clips to change its tint, sharpness, contrast, brightness, or color saturation.

What do some of the more frequently used image characteristics refer to?

Tint: The shade associated with a certain color. For example, a clip might be dominantly brown in color, but with strong hues of gold. The scene would be described as having a gold tint. Some might also describe tint as a tinge of certain colors.

Sharpness: The clarity of visual details in an image or clip. Sharpness might describe how clear or fuzzy a clip is.

Contrast: A measure of the light or color difference that describes a visual object. High-contrast shots include dark areas that are very dark and light areas that are very light.

Brightness: The amount of light either produced or reflected off of a given object.

Color saturation: A measure of how much colors differ from white.

Most programs provide you with standard video filters for changing the appearance of a clip. More complicated programs give you the power to personalize these effects. For example, you can change a clip, originally shot in color, into black and white. How do I do this? Change saturation all the way to zero (removing color and pushing an image towards black and white). Then increase the contrast by 30%. Now if you want this image to seem old, add a little yellow (sometimes called sepia) and increase its granularity. Voila! An image that appears like it came from 1910. Such filters can inject very powerful tones to your video.

Audio Effects

We have focused quite a bit of attention on how you can manipulate the visual qualities of your clips (because this chapter is mostly devoted to graphical edits); however one can also adjust the qualities of sound. Volume is the simplest adjustment.
It is also useful to throw filters over sounds to reduce, say, hiss. Alternatively, you
can add filters that give a track an echo or change the pitch of a speaker's voice.
Similar to video filters, upper-end VEPs give you countless options for personaliz-
ing or microadjusting these effects.

Reposition Clips and Elements

Not only can clips be edited as described above, they can also be repositioned in the
screen framing. Positioning can also be animated. Think of this: You insert a low-
resolution film clip, say, 720 p, into a timeline that is meant to contain clips at a
resolution of 1080 p. Typically, VEPs automatically size the clip within your film
based upon its number of pixels. Thus the lower-resolution clip will not fill your
movie screen. Most viewers would agree that the large black borders surrounding
the smaller clip does not look very good. Luckily, you can increase the scale of the
lower-resolution clip to fit the screen and match the sizes of higher-resolution clips.
Indeed, the resolution of the expanded clip will be less than the rest of the clips, but
at least the scales match. Above and beyond changing scales, clips and shapes can
be maneuvered to different locations around your film screen. This way you can
insert a small-scale video that plays next to a news reporter. You will later learn how
you can use *keyframes* to zoom in, zoom out, or move around a screen as a scene
unfolds. More on that shortly.

Closing Thoughts on Editing

Understanding your options as an editor is the foundation of manifesting your
vision. The true test is taking strategic steps to complete your film. At first, work
from your storyboard to import clips. Then massage the timing and sequencing of
these clips into place for impact. Lastly, it is time to spit shine your work with video
and audio effects and transitions. Magic truly happens through hard work, reviewing your own efforts, and talking to others about ideas or your efforts to date.
Feedback and thoughtful contemplation guide you towards more refined products.
Filmmaking is an iterative process that only improves if you are relaxed and give
yourself plenty of time to accomplish the tasks at hand.

Chapter 15
Useful Video Editing Applications

Use of Photographs in Your Films

From a storytelling viewpoint, sometimes the only way to show a concept is with a still photograph. Let's say that you need an image of the streets of a neighborhood from above. A photo from a NASA satellite will serve the purpose without the need of flying an airplane or drone. Alternatively, maybe you wish to illustrate how bolts, located in hard-to-reach locations, come loose from vibrations. Photos may get in closer and frame small objects, without hand vibrations shaking the camera. High-resolution photos also give you the option to crop in on the smaller objects. Alternatively, historical events may have happened when no one was on hand with a video camera or movie cameras simply did not exist at the time of an unfolding situation. Drawings and black-and-white photographs may be the only way to show certain scenarios and settings, and also embody the art or technology of the time – lending authenticity.

Most viewers have expectations of movies being ever-changing mediums. Perhaps they might even see static photographs, embedded into a scene, as not very engaging, unmoving, and unprofessional. This begs the question, how can you use photographs in movies to complement your content?

Here are some simple ways to increase the tractive qualities of still images in films. These are often-used practices to spark your own thinking about how you can include photos in films:

- Adorn tables and graphs with actively changing elements to draw attention to important regions of the figure. For example, the gradual fade-in of a title or the unveiling of colorful arrows in key locations will draw the eye to sectors of an image where a viewer should look.
- Treat your image as though it is a photo being taken. I already mentioned this tactic when describing the use of a fade-to-white transition to act like a flashbulb. Also try digging through the transitions provided by your VEP to find one that

© Springer International Publishing AG, part of Springer Nature 2018
R. Vachon, *Science Videos*, https://doi.org/10.1007/978-3-319-69512-9_15

Image 15.1 Still images can serve many purposes in filmmaking. Modern VEPs provide you with the means to zoom in on one region of a high-resolution photo and command the program to pan across to another. In this example, panning across the image will gradually carrying the viewer through a mountainscape

looks like the shutter on a camera closing. You may strengthen both transitions with the accompanying sound of a camera shutter.

- Pan across an image. If the resolution of your photograph is large enough, zoom in on one area of the photograph and then gradually move the viewable window across the image. Pan across the image more slowly than you might expect, allowing the viewer to take in gradually unfolding scenes. This tactic was elaborated upon in the filming methods chapter about panning (Image 15.1).

   Panning across an image is a method that was mastered by celebrated documentary film producer Kenneth Burns. For decades, Burns has told very compelling historical tales while leveraging this method to give life to photographs taken at points in history when video cameras were rare or not present. In fact, you might hear this method called the *Ken Burns Effect*.

- Very similar to panning, you can zoom in or out on images. It is important to note – zooming in serves a different purpose than zooming out. Zooming in takes broader concepts and focuses them to something specific. This could be exemplified by an interest in showing safety devices situated about a room. Begin your

Image 15.2 Zooming can be executed using high-resolution photos. The movement of framing, either in or out, on subject matter helps to focus or expand a storyline. Thus zooming steers a viewer's attention. Here we zoom out from a climber to the landscape in which they are immersed

clip with an expansive view of a laboratory and then zoom in on fire extinguishers or alarms. Alternatively, zooming out expands concepts. Let's say that you are talking about water resource managers examining overwatering of certain crops. Begin your scene zoomed in on an individual holding a clipboard and then over the duration of the clip, zoom out to the person standing in an expansive field of corn (Image 15.2).

- Photo collages are warmly received by audiences. Let's say that you took a trip and only snapped still shots, yet you feel the need to retell the voyage in a film. One way to make viewing fun is to treat your images like you are throwing them on a table. Add a white border to each image, such that they may appear more like an old Polaroid instamatic photo – giving the impression that you're sharing printed photographs. Slide photos in from the left, right, bottom, or top, landing them into what appears to be a jumbled pile in the center of the screen. Perhaps add a small spin to the images to elicit the feeling of them spiraling onto the table. What's more, increase the scale of the image when it enters the shot, and then gradually shrink it as it comes close to its point of rest. This strengthens the feeling of the image falling downward onto a table.

Note on Using Other People's Images

If still imagery is one of your solutions, yet you want to use someone else's images, get permission to use the shots. Just like video clips and music, the media of interest might have come from someone else's labors. Honor that and give these individuals the credit that they deserve and likely request. Sometimes people would like payment for usage of outstanding still shots, film clips, or music. If you are strapped for money, try explaining your position. Sometimes people are willing to let payment slide to put their work in front of a new audience or to help a good cause.

Minimum Resolution of Embedded Photos

When embedding images into your film, figure out the resolution of the photographs that you would like to use. Unless you have no other choice, use images that are of equal or greater resolution than what you hope to export for your final video. If you want to export your film at 1080p, then a landscape-oriented photograph must be equal or taller than 1080 pixels and wider than 1920 pixels for the still image to have sufficient resolution. The second that you start to zoom in on photos, resolution issues raise their ugly head. Images with resolution less than that of the film will result in fuzzy corners to elements in the shot or muted colors.

Truth be told, if you are editing for online viewing, you may be able to get away with photos of resolution as low as three-quarters of that of the film that you hope to export. Sure, an audience keyed in on watching for lacking quality will find it, but if used infrequently and appropriately, you can sneak a lower-resolution photo in from time to time. This tactic works best with images that don't have too many hard lines and corners. Such elements show the deleterious side effects of poor resolution first.

You can also distract a viewer's attention from a resolution-lacking image. I have heard a few editors say that audiences are more forgiving of low-resolution images when they are black and white. Maybe it is people's association of black and white to past times when images usually contained more blemishes. See what you think.

A last idea for using low-resolution images – don't let the image fill the screen. Instead, keep it at its truthful resolution (therefore not filling the screen completely) and let the image gradually migrate across the screen. It may add a dynamic touch. Unfortunately, this gig only works if you do the same with other images. Using this technique for one image and not the others cues audiences to something being amiss.

Quality Control and Modifying Imagery

Sometimes you'll find photos that, at first blush, are not very functional for your film. Maybe, the most important element in the shot is hidden in the shadows. Don't simply ignore such images! Instead, take a moment and explore whether simple edits could improve its usefulness. Since you are taking the time to find helpful shots to embed in your film, optimize them (if you are using other people's work, ask permission). Some of the more simple edits include brightening or cropping\ manipulating saturation or sharpness (Image 15.3).

Let's expand this conversation. Not only do simple edits improve a photo, but so might the addition of new elements into the image. How about using text to add the latitude and longitude of a scuttled sea vessel? Or placing an arrow to indicate a tiny fracture along an axle? How about a hoop circling where skin is irritated? These addendums will draw a viewer's eye to information that is otherwise not easy to

Image 15.3 There are many benefits to improving the quality of the still shots that you wish to include in your films. Just like any piece of media included in your film, superior content runs the greatest chance of more profoundly influencing your audience the way that you intend

identify. As a consequence, they can devote more attention to absorbing information outlined by a narration or pondering how the content relates to their own work.

Bottom line: You are making your story easier to follow and for information to be consumed.

The most simple, yet brutish, way to edit and add extra elements to an image is within a photo editing program. Programs like *Windows Photo Gallery* give you some latitude for fine-tuning the light qualities (e.g., brightness, tint, contrast, and saturation) of an image. While you can make very powerful changes to an image in these editing programs, doing too many changes may limit your options down the road. What do I mean by this? First, in most photo editing programs your edits must normally be exported to a separate image file before they are incorporated into your film. Unfortunately, that makes it impossible to execute certain changes to your image once it is inserted in the VEP. For example, in order to move an arrow to a new location, you will have to go back into the photoediting program, execute the changes, and then reexport. What's more, you will have to reinsert the newly edited image back onto your timeline.

Note This condition is not universally true. More advanced video editing suites, like *Adobe Creative Suite*, link their image-editing programs with VEPs. So when changes are made in one program and saved, the changes appear in another.

Let's talk more about issues associated with changing images in photo editing programs. Let's say that you wish to add an image to your film and then actively zoom in on it during a video clip. As you zoom in on fixed images, any elements added in the photo editing program, such as arrows, will grow at the same rate as the rest of the image. One consequence could be losing the meaning in the shot. For example the arrow could fall off the screen completely. It might be more useful to add such elements in your VEP, as opposed to the image editor (Image 15.4).

VEPs are great for editing images when you want your image to act dynamically as a scene unfolds. In other words, VEPs can animate many editing options as a clip plays. For example, you can shift an image from color to black and white over the duration of a scene. VEPs are also great for adding very simple elements such as arrows or titles over an image, where the background might be growing or shrinking, while the elements remain the same scale. Even more compelling, VEPs enable you to animate elements for greater impact. Titles can sequentially appear and fade away as objects are discussed in the image, or an arrow can fly in from off the screen showing the location of a critical feature. Since this is a powerful tool to have in your film producing repertoire, let's dig into animation a little further.

Animated Imagery

The communication of scientific or technical information through film is often complemented by animated graphics. Pictorial representations of relationships between objects are persuasive tools for showing objects at scales very different from what the naked eye normally sees, communicating workflows, or showing how otherwise unimaginable interactions work.

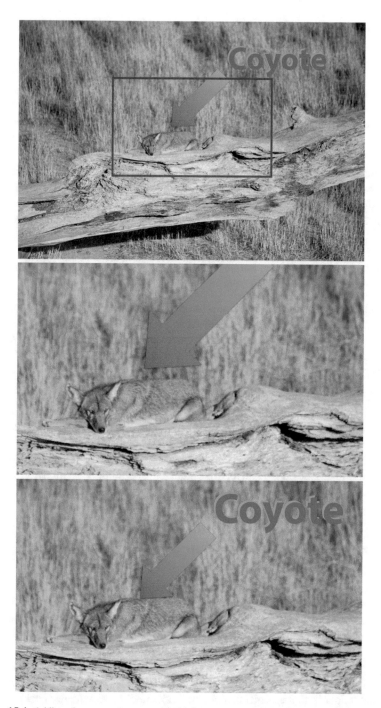

Image 15.4 Adding elements to images, with third-party programs, simplifies the amount of editing that you must do within your VEP. It also can have negative consequences. The first two shots in this series illustrate how a title, well-framed in a wide shot, falls out of the scene when zoomed in. As an alternative, overlaying the title on a VEP track will let you maintain the size and position as you zoom in on a photo

Don't be put off if your animations aren't of the highest production quality compared to blockbuster animated movies. It's likely that the budget and purpose of your film is very different than these blockbusters. Ask yourself, is your goal the precise portrayal of a realistic scenario? Is a convincing backdrop with very realistically moving dinosaurs critical to showing processes like the creation of tar sands from ancient organic matter? How about depicting three-dimensional molecules, when all you want to describe are the valences of electrons? Maybe what you need is a fun and reliable vehicle for describing something that is otherwise hard to envision.

Simple animations add incredible value to films. Very modest animations are effective at explaining or bolstering hard-to-describe concepts. Simplicity may not keep the audience on the hook for long, but can be useful in small doses. What might this look like? A boat moving through a blue sea or a few ideas merging to define how you manufacture a product. What about cancer tumors shrinking within certain regions of a human body following treatment?

Entering into animating, here are a few thoughts to reduce your workload, yet add charisma and appeal to your products.

- Some of the funniest cartoons are oversimplifications or caricatures, and highlight outstanding features or mannerisms.
- Simplicity can highlight important features or patterns. How about graphs that average noisy data signals into a coherent and understandable trend? Or error bars that simplify variability in your signal.
- Think about the generalization of complex machinery or processes. Basic renderings can show how the moving parts of interest operate.

Earlier, we touched upon some very light animations that can be conducted in VEPs; however, now is a very good time to go into greater details. In fact, there is no limit to how deep you can go into animations. Entire classes, degrees, professions, and teams of professionals are dedicated to honing animations. For us, we will continue this discussion on how to animate within your VEP.

Generating Uniquely Shaped Images for Use in Animations

Let's lay down a framework for how to generate descriptive elements, such as arrows or other shapes, that can be embedded and later animated in most VEPs.

For starters, most image files are rectangular. This is true even when you might want to use more specialized shapes (such as an arrow). Many images have the shape, like the arrow, embedded within a background expanding out to give the image it's rectangular shape. In this format, the shapes are not nearly as functional as when they are not restricted to being a rectangle. This begs the question, how do you add uniquely shaped objects (e.g., arrows) without a rectangular background to a scene? (Image 15.5).

Lots of image files are formatted as .jpg or .tiff. While these are very functional formats for many applications, the only shape that they can be is rectangle. As such, you cannot have a .jpg image in the shape of, say, an arrow. However, certain formats, such as .png or .gif files, do allow you to have an image of specialized shape.

A first-order approach to solve this problem is to explore whether the shape that you hope to use, that does not include the rectangular background, can be found on the web in a format. I often look up, say, "arrow, png," on *Google* and then refine my search to view only the image files. It is helpful to know that when you preview images that have transparent backgrounds in *Google*, the background appears to be a white-and-gray checkerboard (Image 15.6). Seeing that is a good omen for its usability. It goes without saying that if such images found on the web are specialized, always gain permission for its use.

What is another option for creating transparent backgrounds to shapes if you can't find an appropriate .png file on the Internet? As discussed in an earlier chapter, let's return to the idea of exporting a file from one type to another (in this case, a .jpg file to .png). This step alone will not suffice, but the idea is on the right track. One must use a suite of editing software that includes image-editing programs like *Adobe Photoshop*. Programs like *Photoshop* give you the option to open the image that contains the element of interest (While it still has the solid background). You can then select regions by color. In the case of *Photoshop*, select the tool called the *magic wand*. If your object of interest is one color, simply select the entire object. If the element that you want is complicated, yet it is surrounded by, say, a black background, you are in luck. In the latter case, select the entire region of black, and then move to the toolbar and navigate to *select* and then *inverse* to change your selection to everything that is not black. Hopefully this now selects the object that you want to isolate. Now copy and paste the selected shape into a new, blank image. The resulting image becomes the shape without a background. Export this new image as a .png file. Voila, this creates the backgroundless image that you desire (Image 15.7).

As a warning, if the background is not one color, parsing out the object of interest necessitates selecting the object by outlining the object either manually or selecting the numerous regions of color that define the object. Both approaches can be time consuming, but get the job done!

You can also make your own shapes and designs with programs such as *Apple Preview*, *Adobe Illustrator* (or *Photoshop*), or *Microsoft PowerPoint*. Most programs give you the option to export figures made in these programs as .png image files with a transparent background.

Adding Elements to a VEP

Descriptive images are added onto video tracks similarly to video clips. If you want more than one element showing up within your frame simultaneously, and you want them to articulate separately, each must be placed on its own video track – one stacked on top of the other. As a reminder, the top clip appears in your film as

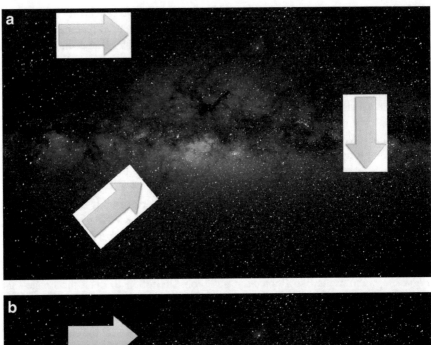

Image 15.5 Sometimes editors want specifically shaped objects to be included in defined regions of a clip. Unfortunately, lots of image files are formatted to a rectangular shape. As such, when you overlap these images over the scene, you might not get the feel for which you are looking. Here we illustrate the inclusion of arrows formatted as .jpg (Image **a**) files versus those that are .png (image **b**). In the former, the white background of the arrows is removed and the impact of the arrows increased

Image 15.6 When
searching the web for
images with transparent
backgrounds, regions of
transparency may be
communicated with gray
and white checkers, similar
to those shown in this
image

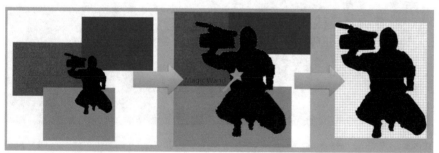

Image 15.7 One technique in *Adobe Photoshop* for removing background elements from an image uses the *magic wand tool*. This lets you select regions that you wish to keep with a magic wand and deselect those that you wish to omit. The subset of selected regions can then be exported into a unique image file

though it is overlaying the clips arranged on lower tracks. As such, you can orchestrate relationships between different elements, giving visual preference to some more than others.

Knowing how tracks relate to the positioning of elements is the foundation for animating elements in VEPs. If you simply place images of the same size, on top of each other, you will only see the topmost layer. By shrinking down the top element,

Image 15.8 When video clips are stacked upon each other in video tracks, they will initially over-lap each other in the center of your resulting film (Image **a**). It is up to you to scale and reposition them for desired impact (Image **b**). Here we moved the clip of a climber to the side of the frame and used an arrow to indicate where the rope is clipped to the rock wall. The scene is comple-mented by a backdrop of gradually gathering clouds

you will begin to see the edges of the next highest layer (Image 15.8). In conjunc-tion, you can relocate, say, the smaller image (or another video clip), to other loca-tions within your screen so that it does not block one that is situated behind it. Picture the example of an interview. The interviewee might be situated to the right side of an image, while you can post supporting imagery to their side (Image 15.9).

How might you animate a couple of simple shapes to help draw the eye to a detail in a background scene? This depends on manipulating the size, angle, transparency,

Image 15.9 Secondary clips can be shrunk down and positioned within the frame to serve as supplemental and descriptive windows

and location of such shapes within the frame. The simplest way to do this is to grab an image file of interest and drag it onto the timeline above your background shot. For discussion purposes, let's say that this is a star shape. All objects dragged onto your timeline will first appear in the center and scaled per their number of pixels.

For most animated scenes the central location won't do. It is up to you to give the element position and life. Changing the position of elements away from the center is different for each program; however, you usually first have to select the image file embedded on a track in your VEP. You then must toggle a control knob that tells you where in your frame the center of the shape shifts towards. Nearby to these levers, you can adjust the scale of the shape.

Now that we know how to position objects within the scene, let's begin to animate them. All intended movements need beginnings and ends. A spaceship moves from point A to point B. You must determine when these moments will happen and communicate that to your VEP. For every change to an effect (such as an element's location, angle, scale, and transparency) you will add a *keyframe*. Keyframing is your signal to your computer to where on your timeline an action or effect will begin, change, or end. As such, the location of keyframes along your timeline define how quickly the element will go from one state to another. So, let's say you want to animate a box. You can tell your VEP to start the box's scale at 50% and grow it to 100%, between times 10:30 and 10:32. You just need to identify the time at which the box is at 50% and 100%. The VEP will automatically progress the size of the box between the keyframes (Image 15.10).

Let's talk about the example of a ball bouncing - found on the first row in the earlier figure. You'll need to first place a *keyframe* where you want the ball to start its movement. Your second keyframe should be where the ball strikes the ground.

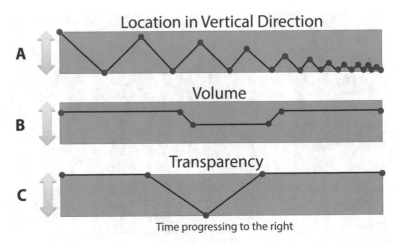

Image 15.10 Keyframing is how VEPs identify where you, the video editor, commands specific qualities of a clip to change. Qualities might refer to transparency, angle, location, or scale (there are countless other qualities as well). Each keyframe represents an inflection point for these qualities. With keyframe points located on your timeline, the program then interpolates the changing condition between keyframes. Here keyframes are used to define (**a**) the height of a gradually slowing bouncing ball; (**b**) a drop in sound volume in, say, a song, so that a speaker's voice can be heard; and (**c**) an object changing to 100% transparent and then returning fully visible. The keyframes used in these examples are represented with red dots

The next keyframe would be the apex of the first bounce while the ball is in the air, etc. To keep the ball bouncing, add more keyframes further down your timeline, identifying the crest of rise and where the ball hits the floor. For discussion purposes, a more accurate depiction of a ball bouncing would include the slowing down of the ball as it reaches its apex of rise. This slowing will either need several keyframes sequenced closely together, such that the ball's rise slows and then begins to descend, gradually accelerating in the downward direction.

For each VEP, the process of placing keyframes varies. Sometimes all that you need do is find the instance on your timeline where you want a condition to begin. Click on the keyframe icon, which may be, say, a clock, beside the controller device. Change the condition to how you want it to be at that point of your timeline. Perhaps changing transparency to 50%. The program will then insert an inflection point in that condition. Next find where you want the changes in the condition to inflect again. Click on the icon, adding a keyframe, and define the state of how you want the condition to be at that point. Maybe 80% transparent. The program will then make gradual changes, or interpolate, between 50 and 80% transparency. When you replay your clip, the animation will change steadily.

Note, once you place keyframes within a timeline, it can be hard to make changes to the locations of those keyframes later in the game. One reason is that a keyframe is isolated to one frame within your film. To navigate to that precise frame can be frustrating unless you turn to a keyframe navigation tool that most programs include. These are shortcuts dedicated to jumping your cursor to the last or next keyframe. Find these shortcuts and use them to save time!

Strategizing for More Complicated Animations

VEPs can easily create movement for a handful of objects; however, building animations with a VEP becomes more cumbersome when more than a handful of objects are autonomously animated. On first principles, VEPs are not designed for complex animations. On second principles, stacking objects into numerous video tracks without clear file identification is clumsy. Objects tower somewhat anonymously on your timeline. Thus, more complicated animations are better built with other programs.

Programs designed specifically for animating allow you to move beyond the very simple movements that can be executed on VEPs so that you can develop more detailed and interactive scenes. When working in such animation programs, organizational strategies save you time, money, and effort. A great place to start for navigating the complications associated with animation is with a plan for how you want objects to look and move within your scene. Here are a few simple guides to help towards these ends.

- Identify how will you represent the elements and characters in your animation. Will they be cartoons, simple shapes, calligraphy, or highly detailed 3-D characters? Since your animations are meant to serve a purpose in your film, the defining qualities of your elements should begin with the needs of your audience. What is their level of understanding? What is their age? What is the need for detail? In general, film producers try to keep a consistent style throughout the course of a film. Too many styles can give a film a cluttered or disorganized feel.
- Conceiving an intricate animation often includes many moving parts. Controlling the number of elements that you want to actuate in each scene to those that are necessary will reduce your workload. Think about which moving parts will communicate what you want, with detail and charisma, but not crush you with a lot of effort. Although the numbers that I mention in the next few questions are not specific, the hope is that they will help you focus your efforts to what is necessary and forgo what may be frivolous. What are 3–5 elements that need to be animated in your scene? What are 3–5 more elements that you would like to animate, but can limit their movement? What are 3–5 elements that are important, but may not require movement at all? What are 3–5 elements that add unnecessary complexities to your animation and thus should be omitted? From there you can decide what scale or level of action will communicate the correct information and evoke the right feel.
- Assemble an animation storyboard. Animations are a complex series of actions that paint a changing picture. As animations unfold, some elements may move simultaneously; however, many also move independently or in sequence. It is important for you to identify the nature of these movements and how they relate to each other in time and space. So that you don't waste time misappropriating movement when you build your animation, grab a pad of paper and turn it into a landscape orientation. Perhaps draw lines to fit the storyboarding template that we shared in the storytelling chapter. If you are going to narrate the scene, write down the words that will accompany the animation at the bottom of each

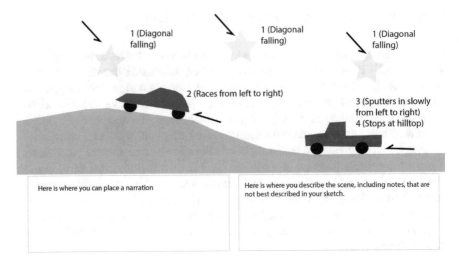

Image 15.11 Simple drawings go a long way to convey how you plan to develop an animation. Use arrows to identify how given elements will be active and notes to define the chronological sequencing of actions

window. Likewise, include any notes that will clarify what you want to happen within the scene. Now use the upper part of your window to make simple sketches of the elements in the approximate location where they belong during that part of a scene. If it makes it easier, label the elements rather than draw them with great detail. Personally, I like to have at least one version of the drawing depict the elements with a fair degree of accuracy to their final portrayal. Now draw arrows where the elements came from or are moving to, and jot down how they get to the next point – say, *spinning*, *accelerating*, or *bouncing*. The difference in location of the elements between each scene will shed more light on to where the elements move. What's more, enumerate which object moves first or which objects move in unison. If they move in unison, the same number can be placed beside a number of objects. Since an object may do several movements as you progress towards your next sheet of paper, leave space for more than one comment (Image 15.11).

Building Animations Without Investing a Dime

Let's continue the conversation about software that you can use to create animations. There are other great options for programs that can produce simple yet effective animations. What's more, many are programs already found on your computer like Apple's *Keynote* or Microsoft's *PowerPoint*. While these are not programs dedicated to animation, they have a couple of notable advantages over VEPs.

- Presentation programs typically don't ask you to place your elements in a timeline. This way you don't have to base your animations off of numerous hard-to-identify tracks stacked upon one another. As we discussed before, working with numerous tracks in VEPs is often hard to wield and navigate. One can more easily manage, say, the animation of a dozen independently moving elements on most presentation-based programs because their workflow emphasizes a user window that shows the physical relationship between elements.
- Presentation programs come equipped with some very specialized transitions between pages. What's more, they also have collections of engaging transitions for the introduction and removal of singular elements into and out of a slide. With a simple click you can have flashbulbs emanate from the contact point between two molecules. Words can bounce onto your screen. Shapes can expand like an accordion. Such unique movements can be hard to rapidly reproduce in VEPs.
- You can record an animated presentation. Most people interact with presentations by clicking a mouse to proceed through the actions or slides. Indeed, you can use this method for creating an animation, whereby you video-capture your screen using such programs as *QuickTime Player*. Press record, go through your presentation, and you will have a film showing the animation that resulted from your presentation. This is not your only option. You can automate the movement and appearance of elements in relation to each other within the presentation programs. For example, one transition can begin 1 s after another action starts. Once happy with your sequence, the presentation can be exported as a video file without the use of screen capture. Unfortunately, these programs usually do not give you the option to export files to a video file larger than 1080p, making its utility in 4K applications a little lacking.

Cool Trick #1

A creative mentality paired with the tools that presentation programs provide go a long way to create effects that most people would never think were possible. Let's say that you are editing a simple animation in *Keynote*. *Keynote* has some catchy introduction or disappearance transitions for objects. Normally, these effects have to be linked to an element, like a box or word. In this way, you can have the box land on your screen, like it is a blazing comet. Wouldn't it be cool if you could just use the transition and not have to include a box in the scene? In other words, wouldn't it be nice to only use the graphics for the transitions without the act being linked to a shape or word? How can you make such effects happen without an associated element? Try adding an element (like a letter or shape) and then introducing it to the desired transition. Next, place the element in the location where you want the effect to manifest. Lastly, make the element transparent. Voila, all that a viewer sees is a very cool transition at a specific location in your frame.

Cool Trick #2
Some programs dedicated to animations are great at creating objects that appear to be three-dimensional. Presentation programs can create relatively effective facsimiles. Think about this: let's say that you have a small box in a scene surrounded by a background of stars. Gradually zoom in on the box, until it fills your screen. Then, command the box to disappear. To a viewer, this measure feels like they passed through the box and out the other side into space.

Cool Trick #3
It is complicated and clunky to build animations or edit images in one program and later embed the results into your VEP. What if you wanted to make changes to your animation and have those edits visible in you VEP. In most cases, you have to reexport your animation or image from one program, and then replace the old animation in your VEP with the new one. This wastes time. Large software companies, such as Adobe, are finding ways to save time and heartache by syncing work from one program with their VEP, *Premiere Pro*. We discussed a similar application earlier when using *Photoshop* to edit an image. If you are planning on doing a lot of such edits, platforms that offer such shortcuts save a heap of time and frustration.

Programs Designed Specifically for Animation

We discussed a number of very simple ways to animate without investing additional money into programs dedicated for the task. However, for small monetary investments you can work with animation programs designed to produce very stylized and exciting effects. Simple cartoons, three-dimensional representations, and morphing of shapes are all within arm's reach with these programs. An example of a program that creates very simple animated cartoons is *Web Cartoon Maker*. A far more powerful cartoon-building program is *Toon Boom*. These programs far surpass VEPs and presentation programs for creating engaging clips, and are specifically designed to keep the building of animations intuitive and as simple as possible.

Don't just grab the first animation program that you find on a *Google* search. If you have a specific style of animation that you want for your efforts, you will be impressed by just how easy it might be to create simple animations that deliver the feel for which you are looking. Beyond seeing what programs can do, these searches open your eyes to new ideas for what you can do with the programs. Fortunately you can use a free version of many programs with enormous animation capabilities. Sign up for a short period of time and learn whether you will get what you need. However, be aware of the trap that when you sign up for the program, billing may start after a week, 30 days, or a certain number of projects. That said, trials are great for determining if the program will fulfill your needs. If you don't like the program, remember to close out your account.

Some animation programs, like *KeyShot, Maya,* or *Adobe After Effects* hold limitless potential for growth and learning. You can devote years to honing your skills with these programs and create amazing animations. To outline such options is as complicated as learning the details of complex VEPs (if not more). Descriptions of these animation programs are found in books. What's more, there is a bounty of knowledge and advice shared in online tutorials and made-from-home video instructions. I find the more time that I spend watching short how-to videos, the more I learn how other people attack certain moviemaking strategies. I catch myself reflecting on how I can apply similar approaches to my own work. All this said, if you want to dedicate yourself to building very complicated and fancy animations, delve into these resources to learn more about your options.

Editing Interviews

Interviews infuse films with human perspectives and personality. They are powerful storytelling vehicles for transporting viewers through logic, turns of events, and the sentiments of people. Some films use them sparsely to strengthen a larger film that might follow a script or carefully crafted storyline. Others leverage interviews as the primary vehicle for carrying entire documentaries. Such films are masterpieces of stitching words, from several interviews, together into a narrative. What's more, the pacing, choice of clips, and supportive media convey authority, fear, confusion, resilience, or resolve. Let's look at some helpful methods for editing films where interviews are your primary storytelling tool.

Single-Person Interviews

We begin the discussion of editing interview-based films with a single-person interview. The simplest solution is to insert the entire interview on a timeline, between title and credits. Add lower thirds to tell the viewer who is speaking and then export to a final film. Simple. Some of us know storytellers who can enrapture an audience for long periods of time. The recording of such people onto film is the premise for *TEDx* talks posted on the Internet. Their personality, reputation, storytelling methodology, and/or content has people wondering what is coming next. This keeps audiences on the hook. Indeed TEDx presentations are practiced, quality controlled, and interspersed with transitions, but they prove that long monologues can provide great content.

What if you don't have a rock star interviewee? Videos that revolve around single-person interviews can still be useful. Many informative films have emerged from simply showing written (or spoken) question on the screen followed directly by answers delivered by an interviewee. In fact, I conducted 50 separate interviews with ocean scientists. Each scientist was asked to define one term, such as *biodiversity*, relate its relevance to natural systems and communities, describe how it is part of the sustainability of the fishing industry, and share how their work is addressing issues around the topic. By the end of the project, we developed a comprehensive online video dictionary for terms related to sustainable oceans. Fifty terms, 50 different videos, all lasting just a few minutes. We found that the short length, compact answers, and conversational tone drove large volumes of traffic to this site.

In order to deliver on gripping short films, it behooves a filmmaker to make some changes to the content. As a first step, you could note where the interviewee strays from a main thought, has a coughing fit, or pauses for long periods of time before answering a question. What's more, you may not want to hear your own questions prompting answers from the subject. This is a great opportunity for you to cut the one-long-interview segment down to its juicier nuggets.

It is easy to get stuck thinking that your film has to play out in the order the interview was shot. This is not true. You can slice and dice the original interview for clarity, coherence, and potency, while preserving the interviewee's original meaning and sticking to your storyboard. Perhaps the interviewee first answered how their product has impacted the automobile industry. It might be more useful to first share the development of the idea that was answered later in the interview and end the film with the product's impact. Resequencing of interview clips is left to the discretion of good storytellers.

When editing segments of interviews, be careful to not lose flow or meaning. For example, the subject may have stammered her way through a discussion about how a product is manufactured. Because of these disruptions you edited out some words and parced together the rest. Unfortunately you might omit valuable content in the process. Thusly, the story may rapidly jump from how the product was designed to its marketing. This is poor storytelling. Leaving the less-than-ideal segment is one option for maintaining the logical progression of your story.

Another solution to dealing with such interview footage includes adding some of your own prompts to guide the viewer towards meaning. Since you started with a title, informing the viewer of the meaning of the film, why not intersperse other titles to inform the viewer about what will be discussed next? Perhaps support the title with a narrator reading the title. Heck, ride the sequence on some appropriate mood-setting music. Should you choose to use this technique, I would suggest applying a consistent format of introducing titles. Inconsistencies, even too much creativity, can provide harmful distractions.

Stories Built Upon Multiple Interviews

Well-constructed films, driven by several interviews, have profound potential to reach deep into a viewer's emotions and motives. These qualities deliver on high potential to impact how they think. Numerous intertwined interviews are the backbone of many documentaries.

How might a story built upon interviews unfold? A first interviewee might introduce a dire condition. Another rephrases the situation to add relevance and share a bit more detail. Yet another interviewee poses questions about how to resolve the issue. The story could then return to the first interviewee who tells how modern science is using a specific line of experimentation to bring clarity to the situation.

The weaving of different people's voices to tell your story shows the faces associated with the field, and lends credibility to the issue. The use of many voices to build a story arc associates each interviewee to the final take-home message. Each person may only talk about one subtheme in the film, but since the film needed those words to build the total pitch, the interviewees are seen as complicit to the film's grander mission. To a viewer, the careful piecing together of numerous voices engenders the feeling that the message of the film is shared, reputable, and thus more largely agreed upon.

Developing a Healthy Mentality for Building an Interview-Based Video

Your mission as the editor of such films is to assemble clips for precision, depth, dynamism, and relevance. Piecing together interviews of numerous individuals to tell a cohesive, even transfixing story, is an enormous challenge. Your charge is to carefully weave the words from numerous people to build a storyline and communicate larger meaning. When relating a story with the words of one person is not easy, cherry-picking words from several people and connecting them in logical progression is monumental.

The process for building a multi-interview film begins with reviewing your clips. The goal is to identify which clips, from all that you have recorded, will add the most value to your film. The prospect of undertaking this task is daunting. While you might have captured 10 h of content, it is common that filmmakers only use 10% or less of the interviewing time within their final film product. Separating the great clips from the chaff takes dedicated time.

I like to motivate myself with perspective on its value. Perhaps this analogy works for you. What if you were stuck in gridlock on a five-lane highway and someone walked up to your car and said, "Take as long as you want, and pick five cars on this road. They are yours, free, without question." Are you going to look out your car window and pick the ones that you immediately see? Personally, I would fill a backpack with water and snacks, and devote my afternoon scouring for exquisite examples of automotive design. Maybe cars are not your thing, but the point stands – if you saw value in what was offered to you, and you knew that greatness is out there, you will go for it with gusto. In terms of interviews, comments that will make your video come to life are out there. They take a while to find. So, go for it with gusto!

During the first pass of interview content, it can be very tempting to dive straight into building one exceptionally strong segment of a story. Unfortunately, diving deeply into this venture before finishing a complete review of all interview clips runs the risk of shortchanging the value all of the content that you have collected. If you have a great idea while reviewing your footage, write it down, even in significant detail, and then return to the review process. Even if you think that you have a topic well-covered, continue with your reviews to flush out whether you have even better comments to carry your story. Sometimes it is tempting to brush over an interview that you felt was subpar and would likely produce few contributions to your story. However, you would not believe how listening to all interviews, including those that might have disappointed you during filming, end up carrying immense potential to invigorate your story. Remember that the review process has a specific purpose – carving out and classifying (into acts) the useful clips from your entire mass of interviews. Once all of your clips are amassed in these acts, the enterprise of placing your clips into gripping sequencing will be better served.

How Do You Start to Build Videos Based upon Numerous Interviews

Reviewing media for your interview is not a time for you to kick back and simply watch clips. Hardly! It's next to impossible to remember all of your reactions and/ or thoughts about where certain segments would best fit Approaching the review process systematically saves time and effort while assuring that you evaluate all of your media. Here are two ways to help reap the greatest rewards from this step. Both are guided by the primary mission of this step – amass all of the necessary interviewee comments that will constitute discrete segments of your story. Our discussion builds upon the previously described methodologies for inserting video footage into a timeline.

First Pass: Classifying Clips

One practical approach for reviewing interviews is to grab a green tea or juice and sit at your desk with a pad of paper. Pull out video file after video file of interview content – press play and scribble copious notes as the video progresses. Never hesitate to press pause or replay clips of interest.

Normally, I take note of valuable comments from interviewees. However, the length of a clip changes the way that the audience processes the message and, thus, the clip's function in your story. I use three categories to describe clips.

1. *Sound bites* may be as short as one word, and rarely extend past 5 s. These are exceedingly short and potent comments that cut to the chase. You might hear, "The action exploded before our eyes!" "What you are about to see really makes a difference."

2. *Golden nuggets* of insight and perspective come in longer segments. Ten- to thirty-second messages are useful at conveying a mood and highlight critical information. Personally, I find *golden nuggets* to be the fuel for a documentary. In 20 s, someone might explain a process or a vivid example. These clips are long enough to maintain the attention of a viewer, yet provide interesting anecdotal information or perspective.

3. Then there are long clips which I call *run-ons*. An interviewee might answer your question, but drag it out, or shift attention to a poor example in the middle of a brilliant *golden nugget* explanation. What could be said in 20 s stretches from 30 s to maybe several minutes. Indeed, *run-ons* can be cut into *golden nuggets*. Conversely, they can be kept as is, yet then you run the chance of boring or frustrating your viewers.

Knowing these categories helps for identifying the utility of an interviewee's comments during the review process.

As you review clips, jot down whether a segment would serve as a sound bite, golden nugget, or run-on. What's more, try to include a few other pieces of information for every note. (1) What is the file name and time code for the comment (e.g., C0003.MP4, 0:01:56)? (2) About what is the person speaking? Since the goal is to identify where clips would fit into the scheme for your entire film, use simple words to identify where these comments would best fit into the workflow of your video. Descriptive comments such as "assembling analyzer," "troubleshooting," or "field testing" will do. (3) What is the quality of the comment? This can be done with stars or simple notes such as "example" or "bad sound," or "expressing frustration." I sometimes classify how much I like them with stars. Note: for extremely poor clips, I opt to save time and skip comments by simply drawing a line through the file name to indicate that it would not add to my film (Image 15.12).

How do you work from these notes to insert the footage onto your timeline? Some folks (myself included) like to work through interviews, taking notes on the clips, and immediately moving the useful clips into place on the timeline of their VEP. Others run through all of their footage in one fell swoop, only taking notes, and saving the movement of clips onto the timeline as a second step. Both approaches have their pros and cons, yet both are best supported by the note-taking process. For example, should you move the clips that you like into place as you review your footage, you run the risk of creating holes in your story. You might not be scanning your storyboard for exact clips, more choosing the ones that you think will fit better in one part of your film over another. Holes in your story will naturally manifest if you are not scanning your storyboard and filling in all of the described scenes.

Image 15.12 An example
of jotting notes while
reviewing interview clips

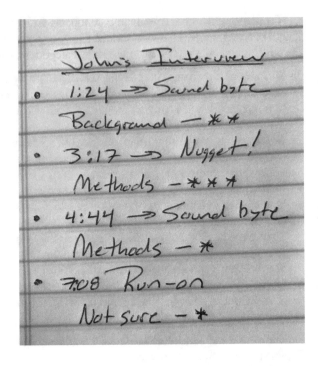

I like to color coordinate clips as I insert them onto a timeline. All clips that are
linked to the theme or act being addressed during given segments of the film are
given a discrete color. Perhaps I choose violet for the introduction, blue for background information, red for fieldwork, green for analytical methods, and yellow for
findings and conclusions (I try to keep a color code decoder beside my keyboard for
frequent reference). That way, as I expand the range of time embodied within my
complete timeline (e.g., an entire 30 min story) I can easily navigate to discrete acts
of my story. At first, these acts may be separated by empty spaces along the timeline. This space gives me plenty of room to insert lots of media into each act. It also
identifies where an act ends and a new one begins. Later, I delete the empty spaces
and coalesce the pieces into a longer masterpiece. Alternatively, more complicated
yet elegant editing platforms give you the choice to develop chapters on separate
timelines or sequences that can later be united in a separate VEP file.

As a first pass, insert more clips than I probably will need. I find it more sensible
to see all of my options amassed together, and then identify and remove superfluous
clips as I see how they relate to the others. Of course, this is a subjective decision.

What if you believe that certain "borderline" clips could add irreplaceable value
to your video, but you are not sure where. Paste these clips at the end of your video
in a separate category relegated for clips without a clear home. That way, when you
start to massage your clips to make more coherent sense, you can devote time to
finding tractive place within your timeline for the homeless clips. Like the color
coding of chapter clips, label these clips with an assigned color so you don't forget
their role.

> **Cool Trick**
> Sometimes I can envision one clip being used at a few different points in my film's timeline. In keeping with the habit of adding all clips that might work to my timeline, I place such clips in all locations where they stand the chance of functioning. What's more, I make notes of all recurring clips so that upon later review, I am reminded to remove all uses except for its best one.

Second Pass: Sequencing for Smooth Storytelling

The binning of your clips, which I am calling your first pass, is a great foundation for staging the next step of building each act of your film. With the first pass executed and all interview clips piled into their appropriate acts, you can focus on sequencing clips for impact. What formed as a bit of a jumble during your first pass is now massaged to tell a continuous, logical, and gripping story. Indeed, we call this the second pass; however, lots of times I end up watching, reviewing, and editing segments many times. So my second pass might actually include the fourth or tenth pass.

A major focus of this period in editing is making certain that all parts of your story are well represented by footage. Flow means that viewers are lead along your desired path with clearly communicated concepts, thus ensuring that you don't lose your audience along the way. Any confusion produced in the proces means that the end of your film is practically guaranteed lackluster impact.

Just the same, too much content detracts from a well-crafted story. Remove any comments that might muddle your message. This might mean that you delete comments that sound good but simply don't contribute useful information. These can feel difficult to delete but in the end this process builds a stronger story.

To increase the impact of my edits during the second pass, I keep in mind a few guiding principles.

First, your ability to tell a story is the Rosetta Stone for sequencing the pile of video clips into logical and gripping progression. Return to your storyboard and refresh yourself of what the string of interview clips is supposed to accomplish. From there, it is easier to identify how to pick the interview clips to serve specific storytelling purposes. It also guides their sequencing for flow. I like to order this flow of themes onto a pad of paper and tape them onto my side of my monitor. Every time that I sit at my work station, I am reminded of how this act is meant to progress.

Perhaps these notes could look like this.

- Learn that there is a problem with old methods.
- Discuss evidence of bad practices (Laura from the shop floor).
- Leave a glimmer of hope that certain methods hold potential for being solved.
- Identify who is scuttling the utility of new methods (dinner party comments on Saturday).
- Launch into how your group went around those who are holding back the field.

Second, ponder the first technique from a different angle. As with any act, you benefit from having a good idea of the disposition of your audience before they start watching. You also should know which state you want them in by the end of the act. Use this information to guide how you piece the interviews together. What do they understand? How should they be feeling at this point in your film? How can you assure that they are feeling this way? Chances are you have written the answers to such questions in your storyboard. If not, reflect on that and add more guiding comments to the notes taped to your monitor. I have felt healthy direction from words as simple as "focus on sharing growing confusion" or "include several short clips reinforcing the fervor."

The state of your audience is largely guided by the qualities of the comments you use to carry your story. For example, who conveys a certain comment lends context. A private citizen discussing the impacts of a war will likely have a different impact than an army general who was calling the shots from afar. What's more, the punch of a comment, like a powerful sound bite or the details of a run-on, communicate very different tones.

Third, listening to many interview clips for coherence is an art. It can be difficult to listen, truly listen, and not be guided by what you want the conversation to convey. Maybe this is akin to conducting robust research. When you know what you want, your interpretation of results may skew in that direction. Bad science happens when teams build experiments with certain outcomes in mind or overexamine data while seeking a specific signal. Thoughts are loaded and steer action towards desired results. Just like looking at data to see what it is telling you, not how it can support your desired outcomes, practice hearing the true voice of an interviewee. Doing so may affect the way that you thread your story together, and will strengthen its authenticity and impact.

Listening to statements from interviewees for authentic meaning opens options. For example, different statements within a long quote can be used to convey different meaning. Think about it, someone could say, "Lots of us were confused, but eventually we became confident that our marketing strategist was on a good track." That statement communicates a sense of hesitation; however, an artful editor could also trim it down to remove the uncertainty: "We became confident that our marketing strategist was on a good track." It may seem like you are taking liberties with an interviewee's words, but the massaging of words within an interview is a long-standing art that has resulted in incredibly powerful messages.

Using words out of context can get you into trouble. Repurposing interviewee statements for purposes other than what they intended to say can stir up a hornets nest of criticism. As an editor, the power and choice to leverage interviews to your ends is within your hands. Ethical editing is up to you. How you use quotes reflects your respect of the person that spoke them. How do I filter my use of statements? If I showed my dentist the raw clip and then how I massaged it, would she raise an eyebrow or just keep picking at plaque? Usually, I would like to think that she would keep on picking.

Cool Trick

The tasteful inclusion of visual transitions between interview segments goes a long way to add impact to the tone of your message. An example? A fade from one person to the next lends a feeling of continuity between people's views and what these people are saying. What's more, transitions can support the pacing and tone that your collection of interviewees' comments is suggesting. If you want to share a litany of criticisms derived from numerous interviewees, a flashbulb transition produces a harsh feel to the conversation. A gradual dip to black brings closure to one conversation so that you can start up a new conversation. It may also happen that you have two clips from one interviewee that you wish to use back to back. What's more, let's say that the two comments don't easily merge from one into the other. There is too much discontinuity in their wording to feel like they seamlessly complement each other. Consider using an abrupt a dynamic transition, like a "spinning cube." It will tell the viewer that something new is coming no matter what the person is saying.

As final thoughts to second passes, it can be difficult to puzzle together wording to communicate all elements necessary to carry a events and logic. With this in mind, be careful to not underrepresent or omit the communication of concepts that must be shared to hold your story together. To do so will increase the potential of losing your audience, and the risk of you not accomplishing the goals of your film. Creating success means that you will have to dig a little deeper or use creative techniques to impart that meaning.

Just the same, it can be easy to reinforce a message with too many comments, thus dragging out that portion of your film. Take time to remove any comments that might muddle your message. This might mean that you delete comments that sound good, but simply don't contribute to your story.

Coalescing Acts into a Complete Video

Your second pass has now led to the development of potent individual acts in a longer movie. These standalone video segments are enormous building blocks in your story. Now you must connect your acts with flow and logic to tell your complete story.

Key Transitional Quotes

Thoughts about how you will transition from one act to another started long ago, with your storyboard. The progression of earlier acts should set up what is to come next. One act, say the background, must be in place in order for you to discuss methods. More specifically, the stage that you set with your background is the vehicle to lend relevance to where your work contributes to a body of work.

However, this is not always the case. Linking acts can be viewed as a third step in the filmmaking process. It is up to you to find or build inter-act transitions. Indeed, these need not be very long, as long as what you use works effectively. For example, you might have a good idea of transitional phrases from your earlier review of interviews. Imagine you found a quote relaying, "after three weeks of document review our research and development team undertook never-tried experimentation." That might be the transition that you need to move from your background act to methods. During your second pass, make sure to insert such clips at the tail end of one chapter to link to the next.

Supplemental Narration

Sometimes it is hard to find interview clips that will carry audiences between the acts of your film. For films based upon interviews, filmmakers have a very practical solution for coupling acts and or concepts: a third-person omniscient narrator. Narrators are guiding voices telling or reinforcing the interpretation of a sequence of interview clips. What's more, the narrator can quickly make very broad leaps of logic, connecting rather obtuse notions. Ultimately, this voice can generate efficient transitions in a film. Simple sentences, such as "The alarm amongst investors prompted a change in research and development," can quickly synthesize what you want the audience to take home from one act and moves them to the next. We go into greater detail on how to record an effective narration in the following chapter.

Supplemental Titles

Titles are another very simple (perhaps overly simple) way to transition the flow of a film. Let's say that your film is moving past the crux into a solution. For example, several interviews may illuminate some dreadful news with no obvious resolution. From your interviewee's comments, you can set up the terrible situation, but you don't have dialog explaining that things changed completely unexpectedly in your favor. How can you say that serendipity saved the day? Consider adding a title saying "Out of nowhere…. the answer" accompanied by some uplifting music. Your next useful interview comment could then pick up explaining the solution.

Titles are finicky. When done hastily, they erode impact, but when done correctly, they are efficient at progressing storylines. Part of their efficiency comes from consistent usage. I would hesitate in switching from narration transitions to title transitions halfway through your film. There is a good chance that this will put off your viewers. Alternatively, introduce the method at the beginning of the film, so that a viewer has appropriate expectations for it's continued use through out your story.

Final Note on Video Editing

Don't lose your labors! VEPs tax computer hard drives and can lead to computers locking up or shutting down. Additionally, editing can progress very rapidly, where countless subtle changes are condensed into short periods of time. It would be terrible to lose your labor. As with all efforts that matter, remember to save your work often.

Chapter 16
Specifics of Sound Editing

Monitoring Sound

Similar to recording sound, audio editing benefits from wearing earphones or earbuds. What's more, one should edit in a room that is otherwise silent. Indeed silence complemented by a brilliant speaker system will also do. If you were to edit simply with a stock speaker system in a room with a whirling air-conditioning unit, you may miss the changes in quality of sound from one clip to the next. Discontinuities and sudden changes in sound quality or volume are dead giveaways of unprofessional or hasty editing. Listening in with high fidelity on what you are manipulating enables you to dig into the subtleties of sound. Need convincing? If you were fishing and wanted to capture as many fish as possible, why wouldn't you use a fish locator? It is a proven tool. This comparison extends into audio editing. Quality listening can be a very inexpensive step towards catching your fish which equates to your film reaching your audience with impact.

Volume Levels Can Make or Break a Video

A filmmaker's sensible use of volume levels can make or break a scene. There are some general rules of thumb to consider as you adjust track or clip levels in your films.

Volume levels should only vary for intended impact. Whether your story is being carried by music, interviews, narrations, sound effects, or ambient sounds, volume levels should peak around the same intensities or decibels as the film advances. For example, many television productions like to keep peak volumes close to and not exceeding ~ −12 dB. This way, as a viewer screens a film, they are not straining to hear or overwhelmed at any point. Likewise it does not bode well when audiences are compelled to turn up or down the volume to enhance their viewing experience. It is your job to create a film that may require an initial volume adjustment by the user at the outset, and that is it.

Rapid volume changes instantly throttle up or down the energy. Know this relationship and avoid changing volume too quickly at any point in your film – unless that is your goal. Let's say you have loud road noise during a car chase, and you want to reduce its volume when the driver of a car is speaking. While your visual transition to seeing the driver might be abrupt, take time to taper the road noise over a second or two before any dialog starts. Not only does it smoothen out the transition, but the gradual taper also cues and readies the viewer for a thematic or action change.

Adjust the volume levels of simultaneously playing soundtracks to emphasize the most important information. Let's say that you hope to have a character speak while carrying a lantern through a forest. If backgrounds sounds, like chirping evening crickets or crackling leaves, are slightly too high, a viewer will lose focus on the words being spoken. This can be agitating. Avoid types of harsh experiences by balancing the primary sound signal by lowering the background sounds. A useful analogy happens for people who like to read the newspaper or conduct work in coffee shops. For some of us, the background energy and sounds enhance our own productive and focused thinking. Unfortunately, one loud and overwhelming voice can ruin the tone that we have come to depend upon, and we can't hear our internal voice. Our connection to our work is lost.

Gaining a sensitivity for the balance of background and primary sounds (like crickets and a character's voice) is difficult as an introductory editor. I think that this is particularly true when background music is meant to support someone speaking. The volume level differences between the music and the spoken voice can depend upon the volume and complexity of the music tracks. A good place to start, with most stock music downloaded off a website, is to drop the volume of the song by a good −20 dB. From there, dial the sound up or down for perfection.

When reviewing the efficacy of audio levels in your film it is natural to be more in touch with the meaning behind a primary sound (like a person speaking) than new viewers. An audiences seeing a video for the first time might not know the value of information being conveyed and may need slightly quieter background music.

To build up an appreciation for these inconsistencies, try taking a good long break away from edited a piece. Then, preview the video with a great set of earphones. Begin the preview from at least a minute before the segment in question plays. This way, you'll get a good feel for the substance and experience that will lead your viewer through the point where you question your sound balancing. How does the sound work for you? Unfortunately, some people find their negative reactions to volume settings immediate, yet positive reactions almost imperceptible. Through this lens, no reaction might be a good reaction to have. Indeed, also pass the video to a collaborator to get some outside feedback.

Editing Sound Quality with Audio Filters

Filters are effective tools for removing some unwanted sounds. Filters work by blocking out sounds bounded within a subset of sound frequencies.

Let's say that you are editing a number of interviews from different geographic locations, sources, or interviewees. One interview (or more) may be quieter than others, thus not having the same sound levels as other interviews. Perhaps the interviewee was distant from the microphone or the person spoke more quietly. Maybe one interview was filmed with a lapel microphone and another with a shotgun microphone. Regardless, sound levels and quality differ. A film's impact on an audience is increased by careful volume and sound quality matching. This may mean increasing the volume levels of one interview by several dB (a unit of loudness). Unfortunately, increasing the volume of a track is usually accompanied by increases in background sounds and/or hiss. These decreases in sound quality are also quite distracting.

Hissing sounds are normally characterized as higher frequency than voices. For example, voices typically reside within the 100–2000 Hz zone, yet hissing may be characterized as above 2000 Hz. VEPs often come outfitted with audio filters that can cut out or reduce the sounds as defined by certain frequency domains. In the case of the hiss, you can keep sounds within the low frequencies and knock out all of those above, say, 2100 Hz. Your hiss problem may be solved! Since this filter blocks out high-frequency sounds and allows low-frequency sounds to pass through, this filter is called a *low-pass filter*. Logically, if you want to block out low-frequency sounds, such as the rumble of thunder, try using a *high-pass filter*.

Audio filters are powerful tools; however, the use of too many filters can render soundtracks lacking in character, if not hollow or dead. I find it advantageous to keep all background sounds unless they create an unwanted distraction or affect the feel that you want your film to have. The key to full and robust sound largely lands on the shoulders of production (quality sound recorded during filming). Manipulations during editing are fixes or tools for improving sound qualities to better- support the tone of your film (Image 16.1).

Adding Music to Your Film

Where Music Fits In

The impact of most films has much to do with their soundtracks. Meaning falters when music doesn't complement the storyline or footage. Have you ever watched a horror movie with the sound muted? How about watching a romance to holiday

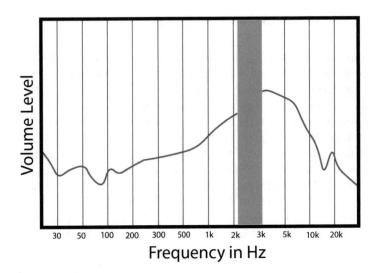

Image 16.1 Audible sounds (to humans, anyway) are constrained to certain frequencies. This illustration shows volume levels captured at different sound frequencies during an instance in a film. Pitch of sound increases from left to right. If unwanted noises land happen within a subdomain of this window, sound filters can reduce their negative qualities. In this figure, a gray box shows where a hypothetical sound filter could eliminate an entire range unwanted sounds

Cool Trick #1

Sometimes sound filters can't fix a soundtrack. Perhaps you can't remove the sound of a lawnmower that plagued the background sound of an interview. The lack in sound quality may become evident as you transition into and out of this clip. The rattling engine intrudes with incredible abruptness, to the point that the words spoken by the interviewee are lost. One solution for dealing with a rough transition into (and out of) a less-than-perfect soundtrack is to ease into the sound with a gradual fade-in (and out). A fade-in gradually increases the volume of the soundtrack of interest. In the case of the lawnmower, consider slowly introducing the rattling sound during the final second or so of the previous clip. This way when the interview clip, plagued by the unwanted sound, is first seen, the viewer is already acclimated to poor sound qualities and ready to hear the interviewee's words.

Cool Trick #2

In the worst situations, when the transition of one clip into another is afflicted by poor sound quality, consider including background music. This will distract attention away from abrupt transitions and unacceptable background sounds. This diversion technique does not necessarily have to be music. Lots of times dynamic visual transitions between scenes can be complemented by an engaging sound. Try using a visual swooping into place of a clip, matched by a *swoosh* sound. Unfortunately, using such transitions to cover a scant few transitions into poor sound is a dead giveaway that you are trying to cover up a shortcoming in your work. As such use such solutions regularly, so as not to draw attention to its application.

music? The souls of the movie are lost without appropriate music. Let's think about how we can use music to enhance the feeling or tone of a film by navigating through stages of films and discussing how songs may come into play.

Introduction

The beginning of a movie is where you can shamelessly draw the viewer into watching the entire film. First impressions are everything. Introductions to films establish a precedent – that you can convincingly wield media to build a story. If you blow it here, it is so hard to draw your audience back in. If you knock it out of the park, you have ignited the storytelling process early. Sure, it would be great to maintain this synergy of content and sound for the remainder of your film, but a great start generates a little wiggle room or tolerance for small disconnects later on in the film.

Music at the beginning of a film is a great way to establish a setting and tone. Lots of films use imagery, including titles, to give the viewer a first glimpse into what the movie is going to be about. Consider how you can use music to bolster whatever setting you hope to establish – a feel of enchanted forests, extreme weather, or bustling cities. What's more, think about what music would launch people into the story that you have in mind, be it adventure, romance, or education.

Body

Music supports an unfolding story. Within most stories is the passing of time, action, and realization. Along with events, storytellers want audiences to journey through specific feelings, and perhaps ignite thought or action within the audience. What are some of the obvious ways that music supports these efforts?

Personal Anecdote

My experience with matching music to movie content began with piecing together ski movies. After a year of skiing the mountains of Colorado and Utah, I would gather the footage that I took and compile the best ski clips of the year. I worked hard to match the spirit of the footage with the progression of great rock songs. When songs hit lulls, I would quiet the volume to include spoken words from the skiers. When songs built, I would emphasize scenes of transition. When tone was rising to a spirited crescendo, I would embed my best action shots. The songs were sure-fired ways to amplify stories and enhanced the moods that I wanted to convey.

I was further propelled to make better ski films with feedback from my friends. Our posse of skiers would have an annual screening at my house. The lighthearted nature of these events was a perfect platform for feedback. I'd quickly learn what whether my song-matching efforts were effective. Months later, in the heat of the following summer, I'd inevitably rewatch my films to rekindle excitement about my next ski season. Incredibly, I'd become a fantastic critic of my own work. The mechanisms that made for blunders and successes in matching film to music stood out like a sore thumb. Still to this day, I sit down and watch my old work with a critical eye. It is amazing how I can use my films from the past to identify what works and what does not work in film (Image 16.2_skiing).

Cool Trick

Most movies begin with music of some sort; however, this is not a given. As a counterpoint to using music during an introduction, crisp ambient sound can be even more gripping. Think of a house cloaked in darkness, with a sole light, emanating from an upstairs window, cutting through the inky black. In the background, you hear an owl hooting and the sound of someone writing with an old typewriter. The simplicity of sound screams someone returning to their old roots, to tell a personal story. Inspiration that comes from deep contemplation is being translated onto paper. Native sounds inject the viewer into a situation.

Titles

Simple technical movies, such as those describing the manufacture process of a new device, are often carried by descriptive titles as simple as "Step 1," "Day 4," or "Chromatography." These titles are a punch in the nose telling viewers what is going to be described next. When done hastily, they may come off as sterile and abrupt, and thus unprofessional. Short musical jingles, like a simple guitar riff, usually

Image 16.2 A clip from one of the author's earlier ski films

decrease this harshness and give the viewer time to assimilate the written information on the screen. I find it generally useful in technical films to accompany titles with sound bites that taper into silence. This step smooths thematic shifts.

Montages

Stories are strengthened by compilations or fusions of clips illustrating a location, process, or concept. We'll call these *montages*. Montages can be just a few seconds to a couple minutes (if you have enough compelling footage). Music is a great vehicle for moving through this content. Try picturing banjo music supporting imagery from a road trip across the United States. How about airy music as a team enters into a vast valley filled with waterfalls and swirling clouds. Or an uplifting anthem as your champion returns from a hard-earned victory? No words are spoken, but the story is advanced through visuals coupled with tonal music.

Background Music

Background music includes musical scores that are easily forgotten, if not unnoticed. Sure they are heard, but they fit so seamlessly into the rest of the media that they are overlooked. They boost the mood that the other, more obvious video and sound clips are attempting to convey. Think of them as the mall music of movies, the caffeine in your cup, or the annoying cold that slows you down – you can sense their presence, but might not always be clear on just how much they influence your experience.

In general, background songs are simple and played at a low volume. They can change character over time, such as reaching a climax; however, these surges are gentler than the louder music that accompanies other scenes, such as montages. Be mindful that distinctive singing in a background song can distract from the storyline or, more specifically, dialog. Audiences may key in on the words of the song rather than the scripted, spoken words that you hope that they'd hear (such as an interview or narration).

Conclusion

A film that is not wound down gracefully can leave an audience left hanging or confused. The last thing that you want is to have your audience walk away from a viewing feeling unsatisfied. Think hard about how you want your audience to feel after their film-watching experience and then choose a closing musical piece to compliment your goal. What might be an example? Songs that end gradually or gently taper in volume ease the audience into reflecting upon how the film's content resonates with them.

Lingering thoughts, particularly if they are favorable, are usually what filmmakers want. Perhaps viewers are tempted into thinking about who they can send the movie link to and share your ideas.

A slow ending is just one way to bring a video to a close. Tones of victory, loss, or doom can be perfect too – it all depends upon your needs. Well-positioned music is elemental in leaving the viewer wanting to see your next movie, engage in a mission, learn more about your topic, or invest in your project.

How Do You Pick Your Scores?

How filmmakers figure out what type of music works for certain situations can seem like a dark art to beginner filmmakers? Here are three guiding thoughts to help you determine which songs will be of use to you.

- The song must suit the needs of your storyline. Think of music as a vehicle for amplifying a tone that you are trying to conjure. Take a moment and note what you need from the musical score. Are you at a point in your film when tension is building to a massive climax? Are team members in transit between locations? Did your team suffer a devastating loss of data because of a power outage? Understanding this allows you to characterize the songs that you need. First, constrain what style of music will be most fitting. Classical? Jazz? Ambient electronica? I then note the mood that you are looking to support, the direction of the mood, the complexity of the musical piece, and the song's approximate duration. For example, "Classical, ominous, building, simple, 90 s." Even when searches are focused, they produce all sorts of results. If you find a song that you like, but are not sure if it fits all of your parameters, consider taking note of the song and then moving on with your search.

- Permissions to use music are a must! Music is an art form that reflects the musician's creativity, passion, and skill. If you are making a movie for your family and friends, permissions are less of an issue, but the second that you put it out where many can see your work, creditless/permissionless use of other people's music is unacceptable. The onus is on you to take the appropriate steps so that you can use a song in your film. This might mean that you pay the songwriter or distributor. Even if use is permitted without payment, it is common that you must give name recognition as you roll final credits. If you can't get permissions, the song is not right for your use - move on.
- Success of the film does not hinge on your approval alone. It is very easy to get into the zone of thinking that the film that you are editing must be guided by your sensibilities. Since the primary goal of your film is likely to impact viewers, consider their needs first. Maybe your ideas for music are great, but verify your choices with people from your intended audience. I also frequently ask friends for feedback on my musical selections.

Song Procurement

Where do you get music for your films? Unless you have large budgets, you likely won't be able to use songs from your favorite album or from the radio. Gaining such permissions is often a long process and/or expensive.

There are quite a lot of songs and jingles (5–20 s snippets of songs) that are free for use. Start by scouring your computer hard drive. For example, should you have a Mac, explore their *iLife* library of jingles. Under such indexes, one can find an basic host of free music and sounds that are very functional in many places within numerous genres of films. Since your choices are not numerous, you will likely find that you hear your music choices played over and over again in other people's work. While the music is appropriate, be aware that other people may have that same experience. The recognition of often-used music typically reduces the perceived value of your film.

In order to expand your choices, take your searches to the web. There you can find countless free resources. Free music, also called "royalty free" or "creative commons" music, can be used with only short attribution to the artist or the online warehouse of music from which you downloaded your selections. While you can vastly increase the number of songs from which you can select for your movies, not all of these scores are top-shelf. If this is the route you plan on taking, be patient and search the web with a discerning ear.

For more choices you can turn to purchasing songs from the Internet. One can search online music warehouses brimming with higher-end songs, with the rights to use as low as $10 per song. Such pieces can add a highly professional feel to your work. What's more, one musical piece may come with a few iterations.

Each iteration is based upon the same song, thus similar, but varies in duration. Differing lengths make each piece more functional at fitting 30 s or 2 min segments in your film.

Don't forget to look for music in your own backyard. Think about great local bands. Whether the band members are your friends or you found a band through a friend, reach out to the band to get a sampling of their music. Don't hesitate to tell them why you might have interest in their work. I have found many bands more than enthusiastic to let me use their music with solely music attribution shown in the credits as your film rolls. In truth, with the web bridging states, countries, and continents, one can often find amateur bands looking to get their music out to the public by any means possible.

You can also become your own music composer. For many, this is an intimidating prospect; however, there are three reasons why making your own music might be your ticket to meeting many of the music needs for your film. First, several computer programs, such as Apple's *GarageBand*, are very easy to learn and come stocked with many useable options. In a few hours, you can learn how to lay down repeating drums and instrumental loops. Stock sound bites can be placed tip to tail resulting in a several minute song. There is another under-appreciated benefit to producing your own music - simple music may fit into your film more effectively than complex. Music produced by a friend's band can be too dynamic for many movie applications, and what comes from your own efforts might better serve your needs. Let's build upon this point. You know exactly what music you need. Like we mentioned above, you might have identified that you need a piece that is "Classical, ominous, building, simple, 90 s." An hour of piecing together loops might manifest just what you need without the need for further permissions.

How Do You Fit a Song into a Scene?

Some introductory filmmakers think that they must use an entire musical track within their film, when in reality many filmmakers only use small segments of songs. I can't tell you how many times I have cropped a song to only use a bit where the music matches the energy of the scene. What's more, songs sometimes dribble to an end, yet I can identify an earlier and better place for it to wind down. Try cropping your chosen song and see how cherry-picking the beginning or end to the song works for you.

It can be very difficult to crop a song to the precise place where you would like it to begin and end without the transition sounding far too abrupt. A simple fading audio transition usually smoothens out these points nicely. These are particularly useful for background music, which is generally more quiet and unassuming than the rest of your footage or sounds.

Sometimes it is easier to manipulate the video content of your film rather than editing louder, more complex, and prominent songs. Instead, let the entire song carry your story. In such a case the music controls the audience's heart, and the visual content serves as secondary media to connect imagery to these feelings. The tone and the way

Cool Trick #1

Musical scores don't need to begin and end at the same time as the video clips that they complement. Try letting the music run past the end of a series of video clips. This way the music merges with the next scene. This elicits a feeling of continuity between scenes. Such a measure is useful when, say, you have background music supporting a conversation between two technicians, but want to transition to contruction of a laboratory. If you taper the background music from the conversation between technicians into your next scene of excavation bulldozers and cranes, viewers will unconsciously connect the conversations scene to the construction.

Cool Trick #2

Opposite to Cool Trick #1, letting a song dwindle and fade a second or two before the end of the video clips can produce some exciting results. In these cases, the sound at the end of a scene is then defined by the native sound in a video clip or even silence. Maybe it's the sound of tools assembling a piece of equipment or feet walking down a hallway. The lack of music can be a potent reminder that the experience portrayed in the video clips has rich and compelling natural sounds. Natural sounds ground the viewer in authentic settings and experiences.

Cool Trick #3

This section of the book is dedicated to giving perspective on how to use music to support the tone of a scene, but don't overuse music. Just as we mentioned in Cool Trick #2, ambient sounds can drive very potent scenes. Try carrying an entire scene without music. This is a very challenging storytelling practice; however, when done effectively, the results are raw and provocative.

that songs progress will more effectively transport your viewer into the emotional space that you may be seeking – and thereby enhance viewer investment. To this end, shorten, elongate, add, or subtract video clips so that the final segment fits the music.

How to Sync Audio Tracks

In Chap. 13 we discussed how to record a concert through a mixer. Sound from all instruments are channeled through the mixer and recorded onto a single track. Differently, some situations benefit from recording sounds, say, from each instrument separately. This way you have fantastic recordings from all sources. Later you can upload each track to your VEP and adjust sound levels there so that each instrument plays at the most optimized levels. This process will require you to synchronize, or "sync," all tracks such that they play in unison.

Let's use and expand this example to shooting a music video. In this scenario you filmed a music concert where separate cameras were devoted to filming the singer, bassist, guitarist, and drummer. Each camera also captured sound through their onboard microphones. For these recordings, your sound was not ideal. However, you also used a portable sound recorder to capture well-balanced "house" sound. This means that you recorded the sound produced by someone who sat at a soundboard, adjusting the sound arriving from all members of the band, and broadcasting the united tracks through the speakers.

In order to make your video of the concert, you want to have the video clips of each musician, synchronized. This way you can switch from one camera to the next, and show the band working in unison. As you transition between musicians, you also need the soundtrack of your video to be solely defined by the track recorded through the "house" sound system.

Syncing audio tracks can be done on your own or with software. Software options are quite intuitive, and advisable if you are going to do numerous syncs.

If you don't have access to synchronization software, a few basic techniques can lead to a more efficient manual syncing. Here is one method. This scenario will be defined such that you will edit a single song, whereby each track begins at some point just before the beginning of the song and ends after the song tails off. If the song were 4 min and 15 s in duration, each camera recording may range from 4:20 s to 4:45 s.

Begin by stacking the video clips of each musician (along with their soundtrack recorded through their onboard microphone) one on top of the other within your VEP. What's more, add the soundtrack from your portable sound recorder. Your timeline will have four video tracks (one for each musician) and five audio tracks (one for each musician and one for the house sound). If recording for each track started at precisely the same moment, you could align the beginning of each clip one over the other. At this time, if you were to press play, you would hear all tracks playing together, and your syncing would be complete. However, it's more likely that the tracks only roughly cover the same period of time, meaning that stacking all tracks such that their beginnings align would result in a preview that sounds like a jumbled mess.

In order to align sounds to sort out the clutter, mute all soundtracks except for the house sound and one musician. This way you can focus on aligning one musician to the house sound without the distraction of other instruments. The key is to find a very short, loud, and distinctive sound that is recorded on both tracks. Although a sound at the beginning of the song makes things easiest, this sound can be found anywhere in the song. For example, you might find a the bass drum beat clearly recorded on both the musician track and the house track. The drumbeat can be your anchor for matching up the sounds of both tracks. For first approximations, some

VEPs give you a visual representation of the volume levels that characterize each audio track. Returning to the bass drum example, you would likely see a spike in sound at every moment that the bass drum plays. So, you can visually sync the tracks by aligning the spikes associated with each hit of the bass drum. Simply drag the musician clips to align with that of the house sound.

If making a visual alignment is too complicating or your VEP does not accurately portay sound bars, you have another effective option. First, take note of the sound qualities for the tracks that you wish to align. Then find a way to identify which track is producing which sound. Hopefully, the house track has higher-quality sound than the individual musician tracks, thus making it distinguishable. If you are listening to two tracks that might have very similar sound qualities, it may be hard to tell which sound comes from which track. In this situation, add a remarkable sound filter or volume change to one track, temporarily. Therefore, if an adulterated sound comes before the same sound on the high-quality track, you now know which directions to move the tracks for them to be synced. Sometimes this means that you shift the audio track by one or two frames at a time to get things just right. There is no getting around the fact that manually syncing soundtracks is tedious.

Cool Trick
Syncing tracks is a very powerful tool, but the process is very labor and attention intensive. If you plan on manually syncing tracks, try to keep the number of clips that you will be moving to a minimum. For example, let's say that you want 10 segments of your music video to show the guitar player. Avoid cutting the long clip of the guitar player down to the exciting scenes before you sync it to the house sound. By doing so you will have to sync all 10 segments individually. Instead, reduce your workload and sync your clips before you slice and dice them down for impact.

With two tracks synced, it is time to move on to the others. The key to repeating this step with the other tracks is to lock in your previous syncing labors. work. Group your two, well-synced tracks together, such that you can move them around your timeline and not mess with their syncing. Next, mute the musician whose sound is now synced with the house sound. Then unmute your next musician's clip. Repeat the syncing steps for each musician. In the end, make sure that the poorly captured sound from all video clips is muted and only the crisp sound recorded by the portable sound recorder is heard. Now the video content of all musicians align and are matched to the impeccable sound. This way, you can cut back and forth from the video clip of the drums to that of the vocalist, as the song plays out (Image 16.3).

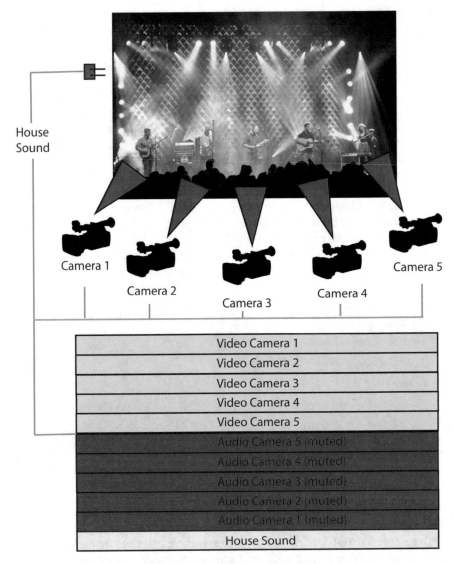

House
Sound

Camera 1

Camera 2

Camera 3

Camera 4

Camera 5

Video Camera 1
Video Camera 2
Video Camera 3
Video Camera 4
Video Camera 5
Audio Camera 5 (muted)
Audio Camera 4 (muted)
Audio Camera 3 (muted)
Audio Camera 2 (muted)
Audio Camera 1 (muted)
House Sound

All Video and Audio Tracks Synchronized

Image 16.3 It is possible to edit together a music video by filming each musician in a band with independent cameras. Once back in your editing studio, you can synchronize the resulting video and audio tracks. This way you can pick the video and audio tracks that you want to display at different points in your film. In this instance, we match the poorly recorded soundtracks (and associated video content) with impeccable sound captured by the "house" sound system. Once synced, the sound recorded by each camera can be muted, and the visuals of each camera matched with the quality house sound. This way you can cut from camera to camera as the song plays out

Final Thoughts on Audio Editing

In order to edit together a film that includes great sound, filmmakers need to use the appropriate recording equipment, techniques for sound capture, and tools for sound editing. Each step along your road to a finished film will either add or take away from the potential impact of your deliverable. As we mentioned much earlier in the book, go the extra mile to execute these final steps with eagerness and dedication. Allow yourself room to grow sensibilities for how you can use different types of sound to relay information or a tone. Sound editing does not come naturally to all filmmakers. However, trust that proficiencies will blossom from practice, learning from films that create a similar outcome to what you intend, and seeking regular feedback.

Chapter 17
Exporting and Distributing Your Film

Film Reviews

It cannot be overstated that feedback on films should be part of everybody's workflow. The amount of time and effort that you and your team have placed into your work disqualifies you from being objective reviewers of your movie. You have massaged ideas for a certain effect; your feelings of scene pacing is deeply connected to your experience in developing the story, filming, and editing. You also know the content that you are trying to communicate intimately. Because of this, you should reach out to others for feedback. You might be surprised by how your film affects others. Even more importantly, the unique feedback that such people can give you could identify a flaw you overlooked. Even better, it might give you a fascinating way to give your work traction.

For easy discussion, I will break reviewers down into two groups: (1) those who are friends or collaborators who did not work on the film and (2) the intended audience outside of your daily circles. I approach both groups a little differently during review sessions. For friends, family, and collaborators, I like to tell them the general gist of the film in advance and my intended audience. Giving them too much information will bias their expectations and may not result in the most productive feedback.

After allowing them time to watch the film, I approach them soon afterwards with a few carefully crafted questions. It is important for them to share their thoughts openly, without too much structured guidance. So my questions are fairly open ended.

"How did you feel after the completion of the film?"

"Do you think that the film accomplished its intended goal (they might have to conjecture about the intended goal)?"

"What were your favorite and least favorite parts of the film?"

"What was the least effective part of the film?"

"What do you wish was part of the film, but was not there?"

© Springer International Publishing AG, part of Springer Nature 2018
R. Vachon, *Science Videos*, https://doi.org/10.1007/978-3-319-69512-9_17

I find that once they have answered some of these more general questions, I can delve more deeply into questions about whether the techniques that I used were are weren't effective at telling different elements of my story.

Differently, for reviews with the intended audience, I like to first identify who these people are before they watch the film. This can be done verbally or formalized by passing out a simple questionnaire. I might ask, "What is your level of interest about the subject?" or "What grade are you in?" This way, I can gain a better perspective on the impacts of the film or the reasons why I might receive certain feedback. A second grader might say that there is too much information, while a 4th grader says there is not enough. And then, I provide the same introduction and questions that I gave to my friends or collaborators.

In summary, feedback is a tool. Other people's thoughts are invaluable and may change your perception of your own films. This could increase confidence in your deliverable or demand changes to tone, information level or flow. If the primary goal is impact within your selected audience, openly accept criticism and see what changes are practical. Some people might not know the financial or production limits that you have on your projects, and may suggest something outlandish. Perhaps they have an alternative solution to a complaint that is unattainable. Own its impossibility, yet hear their wish for another way to communicate a concept or tell your story. Think about more reasonable fixes. Just the same, take each comment with a grain of salt. If one out of ten people does not like your film, yet the others sing its praise, return to the notion that you cannot please everyone.

File Export

You completed your final edits of your film. This version of your film is your *final cut*. Having the final cut of your film saved within your VEP means that through your preview window, you can view all media linked end to end and stacked on top of each other. Your story, as seen through your VEP, is whole! However, unless you plan to have your intended viewer sitting beside you at your workstation or they have the exact same VEP and library of linked media, no one else can screen view your work. Your film needs exporting. Exporting takes your VEP file and all of the media that it joins and compiles the product of your labors into one movie file that most computers can play or can be streamed online.

At this point, it is important to return to the notions surrounding video file formats. Earlier in Chap. 4, we talked about file formats as the container files for the footage, sound, titles, and more that your cameras capture; and that your computer programs use for editing. Editing moves most effortlessly when the video formats of your media align with those that suit your VEP. Differing or numerous formats can lead to slow editing. Similar compatibility issues arise as you hope to broadcast your film into other people's lives. The format of the film that you export largely controls its functionality – not all file formats are created equal. Sure, most computers will play lots of film formats; however, the results will have variable quality and smoothness. There is very relevant logic to exporting using formats that are designed for specific application in mind.

Take a moment to go back to Chap. 4 and refresh yourself about how specific formatting and compressions align with film uses.

Importance of Proper File Formatting

Presentation is everything. What a viewer sees is largely what a viewer experiences. Even if you have edited together the finest work of art, hitches and glitches will change the ways that audiences enjoy their screening. Less-than-ideal playback provokes aggravation, often resulting in apathy, frustration, or indifference to films. What if web streaming got hung up during a very important moment? What if you gave your sister an enormous file to download, but she is a busy businesswoman? A slow download could lead to annoyances that she is going to be late to a meeting. Wouldn't it be better to deliver your film in a format that works best for their needs and limitations that might be set for their screening experience?

The rhyme and reason for appropriate film exporting expands further when the screening is of great importance. Misalignments between what you export and the needs of the screening could be a breaking point in a big deal. Don't shoot your efforts in the foot before people get a chance to see the polished product! Export your films such that your film files account for as many viewer's needs and tripping points in your viewer's screening as possible. In many cases you have enormous control over the format, codec, resolution, method of transmission, and more of your film, so export wisely.

What to Consider When Exporting

What if it is up to you to decide how to broadcast, and thus export your film? There is a good chance that you thought about the precise outlet for your film when you first started storyboarding. If not, now is the time to dig in and specify. Typically you consider two variables: (1) What dissemination platform will reach the largest number of your intended audiences? (2) What is the highest-quality film that you know the viewer will be able to view without complications?

Addressing the first factor benefits from understanding how the population that you want to reach accesses videos. This takes research or insider information. Let's go over some hypothetical examples to warm up your brain. We'll start by identifying the goal of a film or film series, the population that it is designed to reach, and viable dissemination pathways.

- How-to videos directed at 60-something couples looking to improve their skills at salsa and merengue dancing. Research suggests that they will practice after dinner during weekdays. This is an interesting population who might be very hip to web streaming; however, they might also be stuck in the 1990s using DVD players connected to a television. So, your primary avenue of dissemination might be DVDs!
- Five-minute vodcasts (video podcasts) selling global travel destinations to young adults. Youthful minds with the resources to travel extensively usually are quite savvy with the Internet and frequent social media sites. Vodcasts such as these are often housed on *YouTube*, which alone can gain quite a following; however, viewership can take off to untold heights when accompanied by complementary *Instagram, Facebook, Twitter*, and many other social media pathways.
- A 20 min documentary on human resilience in the face of environmental hardship reaching out to couples on a date night. Traveling film festivals are a great way and fun experience for audiences of all ages. Pick the film festivals that you wish to submit your film to, read their film submission guidelines, and follow their directions to the letter!
- A scientific methods video to accompany a sales pitch in a boardroom full of angel investors. Let's make the scenario a little more complicated. The conference is at a resort where Internet is spotty. As a first step post you video on *YouTube*. It will probably broadcast seamlessly. However, why take the chance of failure? If you are going to be presenting off of your own computer, also export it as a high-resolution video file that you can onto your hard drive.
- A network television show. Talk to the network about their needs. Chances are they are going to want the highest-resolution, least-compressed version possible. Follow their directions carefully.

Once you have identified where you will submit or show you videos, formatting easily falls into place. That said, some principles or approaches can save you time and confusion.

- Be aware that frame rates matter. For example, some cinema theaters project film at 24 fps while web broadcasting mandates 30 fps.
- Most online streaming services suggest the H.264 codec to encode your film, with high-definition formats being either 720p or 1080p.
- In general, people watching films on their smartphone don't need film in 4K resolution. Why? High-resolution footage taxes the data plans that they have with their wireless carriers. High resolution also takes more time to transmit. Also, the screen on the phone likely won't play at that resolution anyways. Currently, the highest-resolution screens on phones is about 1080p.
- Moms and dads might dig your film just the same if you send it in 720p resolution versus 4K. Lower resolution might let your mother start viewing easily, and it could save you the hassle of teaching her how to download large files. Oppositely, business partners might want to see the highest-quality export that you can muster.

- Your web speed may be fast; however, you need to consider the web connection of those who will view your video. If the viewers live in a small town in a developing nation where download rates are very slow, you may need to lower the resolution of your films or rely on platforms like *Vimeo* that share both low- and high-resolution versions of films. Full, high-definition footage could be the death null to the viewer that you need and want.
- Does the viewer have time for a long file download before they start watching? Rocket-fast Internet connections are facilitating the faster starting times for streaming video footage, but can you be guaranteed that they have this service? Oppositely, if a client's Internet speed is slow, and streaming is not an option for the screening, send them a link to your file (such as through *Dropbox*) well in advance of the screening. This way you provide them with adequate time to download the file.
- How much time do you have to post the film? Depending upon the file format of a film, the amount of data that each file occupies can vary greatly. Long and high-resolution videos are enormous video files, and thus can take several tens of minutes, to hours, to upload on the Internet.

Hopefully these notions prompt critical thinking about where you might post your videos for viewing, and details of their quality. Internet streaming, video downloads, and DVDs or Blu-ray are a few of the more commonly used dissemination pathways. Let's share a bit more details on their functionalities.

Lots of people lean on web-based hosting, such as *YouTube* or *Vimeo*, to broadcast their videos. Advantageously, these web-based hosts hinge their business on taking your footage and making it functional for its users: the general public. In some ways, they are doing much of the work of film optimization for you! For example, once you upload your video onto their site, some reformat your video into both low- and high-resolution versions of your film. This means that streaming rates can match the Internet speeds of viewers. This way, those with slow Internet speeds can start watching fairly soon after pressing play.

Differently, posting videos housed on *YouTube* or *Vimeo* onto your website may play seamlessly, yet be accompanied by corporate badges and tools that diminish the impact of your film (for their free versions). Additionally, some sites, like the nonprofessional accounts on *Vimeo*, limit you to a certain amount of data space that you can fill every week or month. Unfortunately, one large video may swamp out any options to upload more video content that week. In fact, it is possible that one large video won't even fit. This could wreck your efforts to regularly broadcast new episodes of your show, and thus create a road block.

What if you don't want to use a third-party host like *YouTube*? Now you will have to be more careful with the video formats that you use to post online. Indeed, some folks might think that whatever video file that they post on their website will play as though they uploaded it to *YouTube*. It likely will not. Just like the third-party websites, you have to optimize your films for web viewing. For example, you may post a .MOV file, which does not stream unless it is commanded to. Without taking additional steps to make your film streamable, some viewers may become frustrated with a long download and give up on your video. If you want streaming, it will have to be broadcast from a specialized streaming server on the Internet. Be ready to incorporate such a server into your workflow.

What if you have a client that needs to view the highest-quality video file that you can produce? What's more, you want them to be able to rapidly navigate to various acts of your video without worries that streaming might glitch. As such, you may post your video online for download. In this case, the viewer will not be able to watch the video until it is completely uploaded onto their computer. This removes hitches associated with web speeds from their viewing experience. There are constraints to keep in mind if you house videos on platforms like *Dropbox* (where they can be accessed by other users). First, enormous video files take quite a bit of time to upload. Sometimes, tens of minutes or more, depending upon factors like your Internet speeds, video size, and video length. As such, budget in plenty of time for this process. What's more, uploads occupy Internet bandwidth during the process, and can bog down other web applications that you may wish to run. Consider how you can be forward-thinking about who will be viewing your video. Call them up and let them know that you have uploaded your file to a given location, and warn them that it might take a large amount of time to download. If they need to view the film at a certain time, suggest that they start the download well in advance.

DVDs and Blu-ray are optical disk data storage methods. Blu-ray was designed to surpass DVDs. Such disks were once the most useful ways to pass video files to other users. They could play on home players or serve as containers for passing of data files for your computer. Because of the dual uses, videos could be formatted as either a video file to play in your computer, such as .AVI or .MOV, or as DivX for DVD players. Optical disk data storage is less useful today for several reasons. Simply, fewer people have DVD and Blu-ray players (at home or on their computers). What's more, disseminating your videos means that you ship them through the mail or you must hand them out at conferences. Faster Internet speeds have rendered this means of distribution a hassle. Lastly, if you need to pass on a hard copy of your video to a client try using thumb drives. Solid state memory is far more functional and illustrate a point - before jumping into using optical disks as your method of broadcast, think about whether there are other methods that could better realize your goals.

Marketing Your Work

Delivery of your film to a client is one way to fulfill dissemination responsibilities. This step can be very simple. However, taking on the large-scale dissemination of films necessitates a completely unique set of skills and knowledge. We now have to put on our marketing strategist cap. We have to identify the pathways to reel in audience viewership and then take action.

A great place to start is to identify what approaches to marketing are realistically within your list of options. What could be some examples?

- Paid advertisements on a website or social media outlet.
- Reaching out to friends and collaborators through your personal Facebook page.
- Asking companies, affiliated with your film work, to use their Twitter accounts to promote a screening.
- Passing out DVDs at a conference.
- Renting a booth at a beer festival.

Personal Anecdote
Not long ago, I embarked on a personal film project. The video was not commissioned or sponsored by an organization, nor did anyone come to me with the concept. I conceived of it and grew the idea alone.

It was an experiment for me to convey science to kids in a way that I had never attempted. I harvested footage from years of shooting during travels and set aside personal time to whittle away at the film. I was driven, yet insecure that I could actually accomplish my lofty goal. Progress was made through sweat equity (which is often the case for filmmakers). Slowly but surely, the film took shape.

Finally, the day arrived when I exported the final cut of my film and I had to push my new product onto its intended users. Sure, sending it to a PBS got it in front of people, but I wanted to bring in more traffic than that. Unfortunately, it was such a labor of love, and an experiment, I shied away from using my personal social media platforms to get it out. I believe that I suffered from insecurity that my efforts in a new arena were not enough, so I was shy.

After a month of little movement, I built social media pathways that were branded by the film, not myself. This shifted the way that I perceived the nature of my film. The film had a message that was no longer directly connected to myself. Once this was in place, new traffic surged. Dislocating myself from the message that my work was conveying was critical.

The goal to an effective dissemination strategy is to identify hurdles and find productive ways around them. In this case, I had to get myself out of my own way.

Recognizing all of your options is not always easy. It takes focused and directed thought, and if you have not done it before, you might overlook the obvious. One useful step is to acknowledge the limitations that bound dissemination efforts. Maybe these questions will warm you up to what practical boundaries beset you.

- How much money can you commit to this mission?
- How much time can you commit to this effort
- Do you have any deadlines that must be met?
- Do you have technical or physical roadblocks?
- Does your audience have restrictions to how they can be reached?

From a different angle you can ask, "What are your assets?" Indeed, you might have friends who will print 10,000 DVDs for a fraction of the fair market value. Asset! Maybe you once worked with the director of a local news station who can promote a film festival in which your film is playing. Asset! Alternatively, perhaps your university already has 20,000 followers on Twitter. Asset! Using their social media outlets can help serve your purposes. Most people integrate social media into some aspect of film marketing. It would never hurt for you to post a 30 to 45-second trailer of your video on *Instagram*, *Facebook*, or *Twitter*, so that interested individuals and organizations can quickly get an understanding of what your film is about.

I believe that my best marketing and dissemination assets are usually people. Do you know individuals who have easy access to your viewers? Are there people who have undertaken similar projects that can advise you? Can you tap into their methods and experience? The hope is that you can use what others have honed to reduce the time and frustration that you would expend trying to establish these networks yourself.

In an attempt to unite the last two points, use social media coupled with collaborators or people with similar interests. Reach out to those who have interest in your film and, have greater contact with your intended audience. Perhaps you spotlighted one of their researchers in your film. Don't be shy – explore whether they would be willing to help you out. There is a knock-on effect to reaching out to these people and businesses. It shows commitment of your efforts.

Don't be surprised if 3 months down the road you get a call from one of your partners asking to collaborate again. Perhaps this time, there will be a larger shared pool of money and set of resources for a film or a business proposition.

Screenings

Along with general promotion of your film, it pays to put on a screening. A screening is a dedicated time and location to show your video to an exclusive audience. Whether your 30-min film will be the crowning event of an evening, or your film is sidelight to a larger event, getting people to come together and view your film pays dividends. Word spreads much faster when audiences are caught up in fervor. Maybe the screening will prompt them to ask more questions about your work. Maybe they will use their imagination for where else your film could be screened.

Don't skimp on the event. Share comforts such as coffee or beer with the audience. I have heard people call caffeine "liquid creativity" and beer "liquid courage." Either or both can work in your favor. Also dedicate some time for a question and answer session after the screening. Audiences love when they are engaged in meaningful discussion and interaction.

Lastly, events lend clout to your efforts. Even if people can't make it to your screening, they may pay more attention to your research or products knowing that you're organizing a splendid event or caused quite a stir in the community.

The thought of organizing a screening can come off as an immense amount of effort. Yes, organizing such events can be an enormous burden. One way to lighten the load is to partner with other people or businesses with overlapping interests. Don't be surprised if collaborating meets and beats both of your expectations. Attendance. Viewership. Web traffic. Sales. The list of possible collaborative events are endless. Conferences. Film festivals. Trainings. Wednesday afternoon board of director meetings. Holiday celebrations. What's more, if one event goes off well, other efforts will want part of your excitement. One event can lead to another and another. For projects looking to gain lift-off, producers and causes can find power in numbers.

Final Thoughts on Dissemination

Filmmaking is exhausting. Entering into dissemination is quite different than other steps because you are now leaving the fate of your film in the hands of other people – your audience. Relinquishing control and placing your work in front of people's judging eye can instill anxiety. In turn this can lead to apathy. Just like any race, finish your filmmaking efforts with strength and a positive attitude. All of your work culminates in your film reaching the people that came to mind while you were crafting your storyboard. So do it! Effective dissemination stems from strategically uniting contacts, time, and money. You have been smart so far, so let your labors be realized. Behind every option to screen or distribute your film lies potential for landslide reactions from audiences. Take strident steps to let your dreams of people enjoying, marinating in, or gloating about your film happen.

Epilogue

Filmmaking is a wondrous harmonization of science and art. Ths is why I love it. Filmmaking certainly benefits from natural inclinations. However, it also draws on so many different skills – creativity, thoughtful planning, vision, determination, open-mindedness, communication, technological understanding, and logic. Every day in the process is different and how you combine these tools is constantly evolving. Movies take time to develop and growth happens in uneven steps. It hurts. It surprises. It excites.

Everyone needs a starting point from which to develop their skills, though. I hope this book provided you with that foundation and the inspiration to take filmmaking more seriously. You will learn that the methods that you use have as much to do with the equipment and programs as your interest level. Be ready to mature and learn in ways that this book and others can't portray. In fact, I suggest that sometime in the future, return to the resources (like this book!) from which you blossomed. Your own experiences will cast different meaning on the concepts, possibly resulting in new perspectives and insights.

What's more, go back and review your early work. Let's say that you create a movie this year. Review that movie next year or in 5 years. You may be surprised about how innovative or how skilled you were at storytelling. Alternatively, you may recognize that you have grown quite a bit since then. Strangely, seeing the successes and mistakes that you produced years ago can have more influence on how you approach your next filmmaking enterprise than any book or other person's film.

Remember to be plastic in your growth.

It is easy to come to depend upon a certain tool or style of film. Why? Because you are good at them and learning is hard. Those tools can be your signature. Professionals, however, are always stronger when they have more than just a few techniques. Never stop talking with other film producers. Listen to reviews of your own films with open ears. Focus on hearing what the reviewers have to say and try to avoid emotional responses when you don't sweep people off their feet. Just the same, you will learn about advancing technologies. More and more often, you will find that you are coming from a background based on experience. You will have

© Springer International Publishing AG, part of Springer Nature 2018
R. Vachon, *Science Videos*, https://doi.org/10.1007/978-3-319-69512-9

opinions about whether these methods will complement the work that is already under your belt. You will pass on your knowledge, perspective, and advice to others. This is exciting.

Lastly, as filmmakers, our work is a popular and very powerful to influence other people's lives. We inspire people. We stir up changes in thinking. The way that we storytell well-known facts can alter the status quo. Films steer the way that people perceive their future or see their place in the world. These are not small effects. Not everyone has such powerful levers and pulleys in society. Imagery, music, and the power of storytelling command people's hearts and minds, which often results in action. Go see what you can do!

Index